International Conference on Ecology and Biodiversity Across Time and Space (ICEBATS 2018)

IOP Conference Series: Earth and Environmental Science
Volume 380

Penang Island, Malaysia
15 - 16 August 2018

Editor:

Shuhaida Shuib

ISBN: 978-1-7138-1300-2
ISSN: 1755-1307

Printed from e-media with permission by:

Curran Associates, Inc.
57 Morehouse Lane
Red Hook, NY 12571

Some format issues inherent in the e-media version may also appear in this print version.

Institute of Physics
Dirac House, Temple Back
Bristol BS1 6BE UK

Phone: 44 1 17 929 7481
Fax: 44 1 17 920 0979

techtracking@iop.org

Additional copies of this publication are available from:

Curran Associates, Inc.
57 Morehouse Lane
Red Hook, NY 12571 USA
Phone: 845-758-0400
Fax: 845-758-2633
Email: curran@proceedings.com
Web: www.proceedings.com

TABLE OF CONTENTS

IOP Conf. Series: Earth and Environmental Science **380** (2019) 011001 doi:10.1088/1755-1315/380/1/011001

Preface

The first International Conference: Ecology and Biodiversity Across Time and Space (ICEBATS) 2018 was held at the School of Biological Sciences, Universiti Sains Malaysia, Penang from the 15th to 16th August 2018. This conference encompassed a joint effort among Universiti Sains Malaysia, SEAL NET, and LIPI to address critical issues throughout the South East Asian regions particularly large scale anthropogenic activities that are unsustainable and could impact ecosystem function, biodiversity and the well-being of human populations. This conference was attended by 41 registered participants and served as a platform for researchers, scientists, managers, policymakers and local communities determined to find sustainable solutions to preserve natural resources and biodiversity of this unique region. As a result of diverse research backgrounds, we believe this conference proceeding covers many topics of significance for the reader. Contributions from authors feature findings from fields of ecology, biodiversity, mathematical and ecological modelling, technological and engineering approaches to resource and disaster management, sustainable potential of biocompounds, and human health among others. It is with hope that this proceeding will strengthen the basis for further research, cooperation and collaboration among various scientific disciplines in an effort to enhance the conservation and preservation of our natural resources and ecosystems in the South East Asian region.

Conference participants, organizing committee and invited keynote speakers pose for a group photo on the first day of the ICEBATS 2018 conference at the School of Biological Sciences, Universiti Sains Malaysia.

Presentations from attendants brought forth findings that are vital towards understanding the importance of preserving biodiversity with sustainable management practices in various aquatic and terrestrial ecosystems in the ASEAN region.

IOP Conf. Series: Earth and Environmental Science **380** (2019) 011001 doi:10.1088/1755-1315/380/1/011001

Conference participants pose in front of the ICEBATS 2018 poster with newfound colleagues.

Conference participants gather during the welcoming dinner held at Anjung Budi USM after Day 1 of the ICEBATS 2018 conference sessions.

Day 2 of the ICEBATS 2018 conference dinner at the Top in Komtar, Penang attended by invited keynote speakers and the organizing committee.

We wish to thank, and acknowledge the contributions from the following organizers and co-organizers:

Southeast Asian SEALNet Limnological Network

Akademi Sains Malaysia

50 Years of Leading Knowledge for Change USM; SBS School of Biological Sciences

Malaysian Scientific Association

LIPI Lembaga Ilmu Pengetahuan Indonesia (Indonesian Institute of Sciences)

Young Scientists Network Malaysia

Kementerian Tenaga, Sains, Teknologi, Alam Sekitar dan Perubahan Iklim

penang convention & exhibition bureau

Life Tech Solution Venture

Warm regards,
Shuhaida Shuib
ICEBATS 2018 Editor
School of Biological Sciences
Universiti Sains Malaysia

IOP Conf. Series: Earth and Environmental Science **380** (2019) 011002 doi:10.1088/1755-1315/380/1/011002

Peer review statement

All papers published in this volume of *IOP Conference Series: Earth and Environmental Science* have been peer reviewed through processes administered by the proceedings Editors. Reviews were conducted by expert referees to the professional and scientific standards expected of a proceedings journal published by IOP Publishing.

IOP Conf. Series: Earth and Environmental Science **380** (2019) 012001 doi:10.1088/1755-1315/380/1/012001

Climatological and land use evaluation to maintain water supply in Pulai reservoir, Bintan island - Indonesia

M. Fakhrudin[1*], Luki Subehi[1], Hendro Wibowo[1]

[1]Research Centre for Limnology, Indonesian Institute of Sciences, Cibinong 16911, Indonesia.

**Corresponding author: mfakhrudin@limnologi.lipi.go.id*

Abstract: Pulai reservoir with an area of 49.9 ha and maximum depth of 12 m, serves as clean water resources for Tanjung Pinang city, Bintan island. The objective of this research was to evaluate the fluctuation of water level on reservoir, related to the climatological and land use conditions. Water level in January 2009 was 400 cm and continued to decrease to 105 cm in August 2009 that ceased the water supply. The rainfall during ten months study period (January – October 2009) was 1,968 mm, below of its average level of 2,447 mm. The dry spell reduced the water supply input to reservoir for about 2.5 million m^3 or equivalent to four months of clean water supply. Mean monthly air temperature was also increased by 1 °C compared to the last 29 years. A similar trend was observed for humidity, where its mean value of 85.9% decreased to 83.3%. Initially, the catchment area was 474.12 ha, covered by protected forest (98%). However, in 2009 the remaining forest area was only 8.15%. It suggested the decreasing of water storage capacity. It could be explained that the climate change and uncontrolled land use may contribute to the water crisis of Pulai reservoir.

1. Introduction

Watershed is a stretch of area which is limited by topographic boundaries (ridge), which if the rain is to be accommodated, flowed through the tributaries and out on the main river and then to the sea. This changes rainwater into the river flow and flow into the sea through a long process in the watershed. Therefore, the watershed can be assumed as a hydrological system, where rain is an input and the characteristics of water product (quality, quantity and distribution of water) flows into the sea as an output. If the process is still running well, the surface water fluctuations and sediment content are relatively smaller [1].

The process is influenced by hydrological, geomorphological, geological, topographic, climatological, soil and land use factors. These factors are interrelated with each other and land use is a factor that is rapidly changing in accordance with the development of population and socio-economic level of the community.

Land use change has a linear relationship with an increase in built area. In forests, the soil surface is mostly filled with litter that hold rainwater flows, slowing down surface flow and can further increase soil organic matter, so that the soil is looser, microorganisms thrive, all of which will increase infiltration capacity and permeability of soil. In addition, plant roots also increase infiltration capacity and water permeability.

In the conservation context of natural resources, according to [2] plants have many functions. Plants intercept rainwater which reduce rainwater energy (splash erosion) and reduce the runoff, decrease the runoff velocity, decrease soil displacement, increase soil aggregation and porosity, increase biological activity in soil and soil moisture and also increase storage capacity.

Land use changes affect the infiltration capacity and surface water storage or a combination from both, and the subsequent effect is to cause changes in surface flow [3]. The decrease in infiltration capacity has more influence on the volume of surface flow, while the surface reservoir has more influence on the slowing (delay) of the surface flow to flow up to the outlet watershed.

In addition, for land use factors, rainfall is a very important factor in the watershed hydrological system. Rainfall patterns affect the processes that occur in the watershed. If the characteristics of the rain change, the yield of the water coming out in the watershed will also change. Global climate change causes a shift in the season, where the dry season will last longer resulting in a prolonged drought.

According to [4] the impacts of global climate change, increases of extreme events, such as huge rainfall, high air temperatures and storms, could also cause water scarcity. This water scarcity has a much greater intensity and impact on small islands. In small islands, the river flow is short, the catchment area is small, thus water reserves both on the surface and below the ground are very limited. When there is only a small disturbance to rainfall, the impact on water reserves is strongly observed.

Indonesia is the largest archipelagic country in the world which consists of 17,508 islands that have a very high level of vulnerability to disasters and climate change. Climate change exacerbates and increases the complexity of problems in small islands such as the availability of clean water, sea water intrusion, sea water abrasion and ecosystem degradation.

Bintan Island is one of 111 outermost small islands that have been set by the government based on Presidential Decree No. 6 of 2017 concerning the determination of the outermost small islands. Bintan Island is a specific national strategic region, with natural resources and environmental services. It also has a strategic role in maintaining the sovereignty of the state. Bintan Island, Batam and Karimun are planned as a development area for Special Economic Zones (SEZ) through collaboration with the government of Singapore. This SEZ is a node of the centre of superior economic activity, supported by excellent service facilities and the capacity of infrastructure that is internationally competitive. The development of this special economic zone certainly requires large amounts of water but on the other hand, the availability of water is very limited. In the future, pressures on water resources will increase, both from the impact of economic development, such as land conversion, pollution and over exploitation; and the impact of global climate change, such as an increase in extreme rain.

Pulai Reservoir lies in Bintan Island as a water supply for Tanjung Pinang city. Based on the interpretation of the 2007 SPOT image, the reservoir area is approximately 49.9 ha. The maximum depth of the Pulai reservoir is around 12 m with a relatively flat reservoir. This reservoir is elongated with four rivers flowing into the reservoir and has a spill way at a height of 11.75 m.

In 2009 Tanjung Pinang city, experienced a water crisis, which was caused by the occurrence of extreme dry season. Pulai reservoir, which is the main source of raw water for PDAM (national drinking water company) Tirta Janggi, experienced a significant deficit, almost three months from June to August that influenced the operational PDAM. Therefore, the objective of this research was to evaluate the fluctuation of water levels in the Pulai reservoir, related to climatological and land use conditions, and how to increase the capacity of the Pulai reservoir.

2. Materials and methods

The data in this study was obtained from secondary data and observations in the field. Secondary data, such as rainfall and meteorological data was obtained from Tanjung Pinang meteorological station which is located at an altitude of 37.5 m above sea level [5]. Reservoir water level and intake discharge data, as well as water production capacity were collected from PDAM Tirta Janggi Tanjung Pinang, and the protected forest area map was obtained from the Tanjung Pinang forestry service, as well as data and information related to the Pulai reservoir from the Riau Islands government [6]. Data on land use types was obtained from the interpretation of satellite images and field ground check.

2.1. Processing and analysis of data

Physical data processing of study areas related to aspects of space were analysed by Geographic Information System (GIS), including distribution of land use, topographic distribution and determination of the catchment area of the Pulai Reservoir. Meteorological data processing was carried out with weather-parameter statistical analysis of data series for the past 30 years (1980 - 2009).

The availability of water in the Pulai Reservoir was based on the calculation of monthly flow derived from the catchment using the Thornthwaite and Matter water balance methods. This method requires input of monthly rainfall data, temperature, astronomical location, land use map, land map and average height of watershed. Land use maps and land maps were used to calculate "Water Holding Capacity (WHC)", which is influenced by soil texture and root zone. Based on the analysis of these two param, the WHC value of the study area was estimated at 150.

The distribution and accumulation of monthly flows from catchment were used for the analysis of reservoir water balance. Cumulative flow was analysed using the "Simple Mass Curve" approach. The mass flow rate of the incoming flow can be used to determine the amount of water available that is safe to be used throughout the year. The amount of availability of water was determined based on the cumulative flow debit slope [7]. The water balance of the Pulai Reservoir was calculated by considering the incoming water from both of surface flow and rainfall that flows directly on the reservoir; and the water taken for the interests of PDAM and evaporation directly from the reservoir.

3. Results and discussion

Analysis of reservoir water level data from 2008 to 2009 showed that during 2008, water fluctuations ranged from 270 to 400 cm (the height of the PDAM intake at the position of 0 cm). In the first week of January 2009, the reservoir water level was in the position of 400 cm, but the second week immediately fell 100 cm, to 300 cm and dropped continuously to -1.05 cm (fourth week of August). In the third week of September the water level was at 68 cm and increased steadily, in November the reservoir water level was already 205 cm (figure 1). During the 3 months (June-July-Aug) 2009 the supply of water to the PDAM was severely disrupted, so that the supply of clean water to the city of Tanjung Pinang was reduced by around 50%. The lack of raw water supply in the future is increasing considering the increase in PDAM intake, in 2006 the average volume of water per month was 510,986 m^3, in 2007 it was 541,771 m^3 and 2008 was 573,613 m^3 [8].

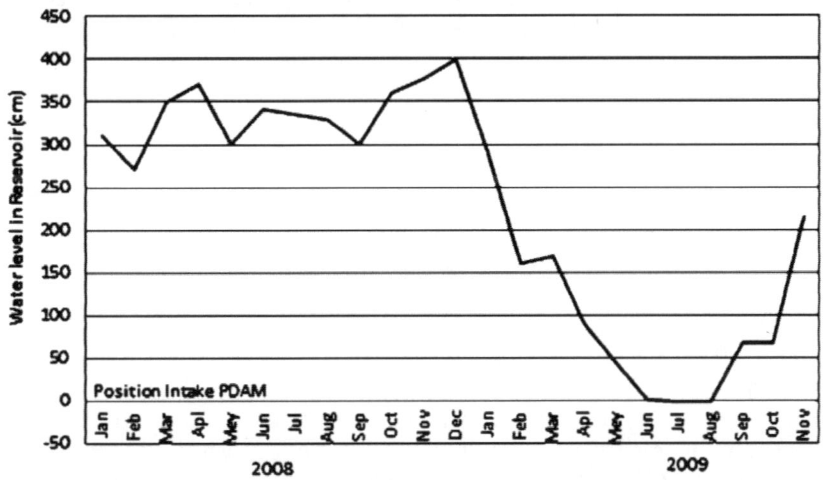

Figure 1. Water level fluctuation at Pulai reservoir.

3.1. Climatological analysis

Rainfall and evaporation are factors that influence the dynamics of water resources in a watershed. This evaluation focused on the current state of water crisis in Pulai reservoir in 2009 and compared to the average conditions during the last 29 years, using data sourced from [5].

Rainfall data shows that the average annual rainfall from the last 29 years (1980 - 2008) was 3,216 mm. Over the last 29 years, there has been a dry year twice, where the annual rainfall was well below the annual average, which amounted for 2,417 mm and 2,396 mm for 1997 and 2000, respectively. The condition during the water crisis in 2009 was a repetition of the phenomenon of dry years. Meanwhile, precipitation during rainy days in the study area recorded fairly large amount, with the average of 196 days per year. This condition can increase the moisture, so it can reduce the rate of evaporation either of the reservoir and the water catchment area of Pulai reservoir.

Rainfall in 2009 was smaller compared to the average of the last 29 years (figure 2). During the period of January to July 2009, the amount of monthly rainfall was smaller than the average during the years 1980 to 2008 (except February), even in January amounted to only 42 mm, whereas the average of the last 29 years amounted to 301 mm. Similarly, the amount of rainfall during 2009 was 2,563 mm, below the average in the last 29 years, which was 3,216 mm. The decreased rainfall of 653 mm reduced the supply of water to the reservoir for about 20 percent of the average rainfall during normal conditions.

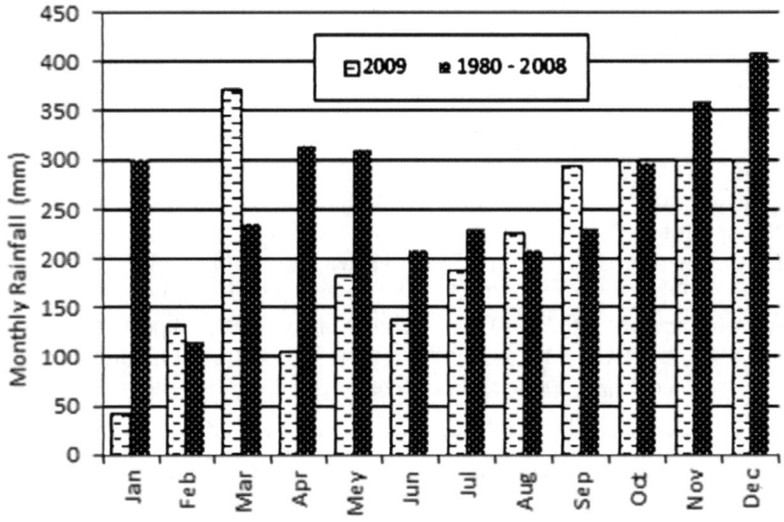

Figure 2. Comparison of rainfall in 2009 with the mean values of 1980-2008.

Climatological elements that affect evaporation consist of wind speed, humidity, solar radiation and air temperature. In 2009, air temperature was higher than the average for 29 years (1980-2008) and on average, it increased by around 1 °C. The decrease in water content in the air (humidity) occurred in almost all months in 2009 compared to the average of the last 29 years, except in February, it was slightly smaller (80.7% to 81.1%). Air humidity decreased over the year from 85.9% (in 1980 - 2008) to 83.3% (in 2009).

Based on the conditions of the two elements (air temperature and humidity), in 2009 the Pulai reservoir and its catchment area experienced an increase in the amount of water "lost" through evaporation. Meanwhile, the rate of evaporation was relatively disturbed by solar radiation and wind speed. Overall, solar radiation in 2009 decreased compared to the average of 1980 - 2008, from 46.5% in 1980 - 2008 to 39.5% in 2009. The declining pattern of wind speed was also observed, which was from 7 knots in 1980 - 2008 to 6.8 knots in 2009 [8].

3.2. Land use analysis

Land use is a rapidly changing factor in accordance with the development of population and socio-economic level of a community. This land change affects the hydrological balance of a watershed.

The water catchment area of the Pulai reservoir (530.4 ha) was actually legally protected, because almost all (98%) of the reservoir water catchment area was covered by the protected forest area (figure 3). This Protected Forest Area has been set since 1987 (Minister of Forestry Decree No. 424 / Kpts-II / 1987, December 28, 1987) when Tanjung Pinang region was still relatively undeveloped. On the other hand, the development of city and population influenced land use and reduced forest condition.

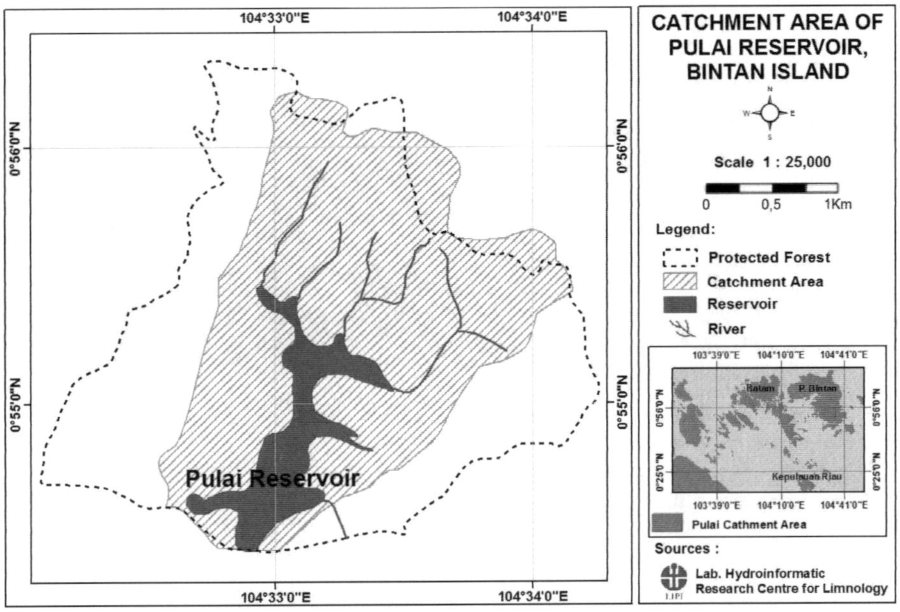

Figure 3. Catchment area of Pulai reservoir.

Land use in the Pulai reservoir watershed based on the data series of 1990, 2003 and 2007 is presented in figure 4. In 2007 the forest area remained only 8.15% and the other in the form of shrubs 50.8%, plantations 36.7%, mixed gardens 3.5% and settlement 0.8%.

Forest area decreased drastically between 1990 and 2003, from 65.97% to 4.65%, although in 2007 the forest area increased to 9.15% but not significant. Meanwhile, on forest land, the surface of the soil was mostly filled with litter which hold the raindrop, slowing down the surface flow, this process further increase soil organic matter, thus the soil became looser, microorganisms thrive, all of which increase the infiltration capacity as well as soil permeability. In addition, plant roots also increase infiltration capacity and water permeability.

Deforestation and changed forest into open land until 2003 was recorded at 52% of the total land area of the Pulai watershed and changed into shrubs in 2007 (51%). The area of shrubs which covered more than half of the Pulai watershed were potentially turn into agricultural land or other uses. Considering that all of these water catchment areas are included in the Protected Forest area, it is very possible that this stretch of shrub can be reforested in order to increase the carrying capacity of the land for the sustainability of the Pulai reservoir. Forest land clearing was related to the development of plantation areas, where in 1990 was only 18.52% of the total area of the Pulai watershed and increased to 41.66% in 2003 [9].

Land use in the water catchment area of Pulai reservoir in the form of shrubs was more than half. Reforestation will increase the water reserves in the reservoir catchments. According to [10] coefficient runoff on shrub (0.07) was two times higher than its forest (0.03), indicating more surface runoff in shrub. Also, it could be explained that infiltration capacity in forest was larger than that in shrub.

The percentage of rainwater that seeps into the soil is very important in the conservation of water resources. Rainwater that seeps into the soil can then flow into the part that is saturated with water as groundwater supply. During dry season, the saturated water also stabilizes the river water flow, as a base flow at a small discharge [11].

Forests function as an effective rainwater catchment area. The soil layer is very rich with organic materials, so the infiltration capacity becomes higher. The greater percentage of rainwater that seeps at the surface of the ground, it will reduce the risk of flooding, drought and sedimentation, and decrease soil erosion [12].

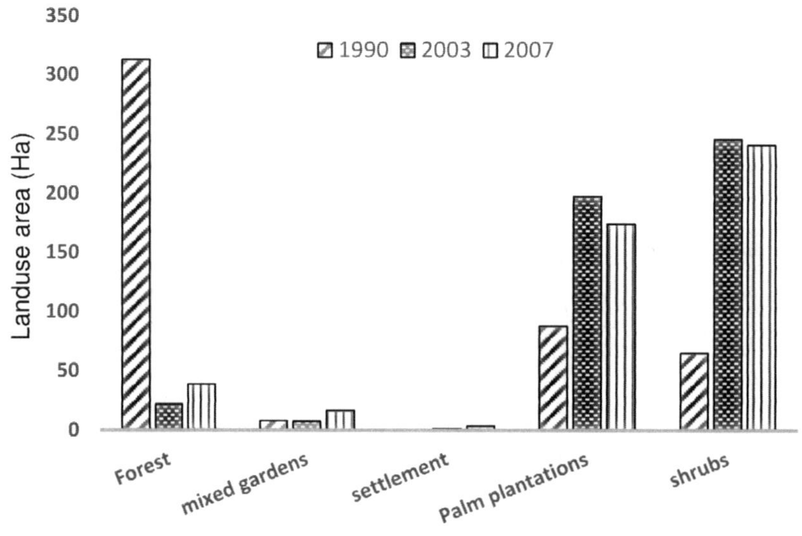

Figure 4. Land use of Pulai reservoir catchment area.

3.3. Water supply condition at Pulai reservoir

The availability of water in the Pulai reservoir was calculated as flow, derived from the catchment using the Thornthwaite and Matter water balance methods. Flow distribution and accumulation were used to analyse reservoir water balance. Cumulative flow was analysed using the "Simple Mass Curve" approach. The mass flow rate of the incoming flow can be used to determine the amount of water available that is safe to be used throughout the year. The amount of water availability was determined based on the cumulative flow rate slope. The accumulation of flow that occurred from the catchment area of the Pulai reservoir is shown in figure 5. The figure shows the water discharge limit for PDAMs with a monthly volume of up to 475,000 m^3 without changing the water level in the reservoir. If the intake of water is greater than the discharge, it will reduce the volume of water in the reservoir. In addition, if the rainfall is lower than normal, the water volume will continue to decrease to a level lower than the intake point, as happened in 2009.

IOP Conf. Series: Earth and Environmental Science **380** (2019) 012001 doi:10.1088/1755-1315/380/1/012001

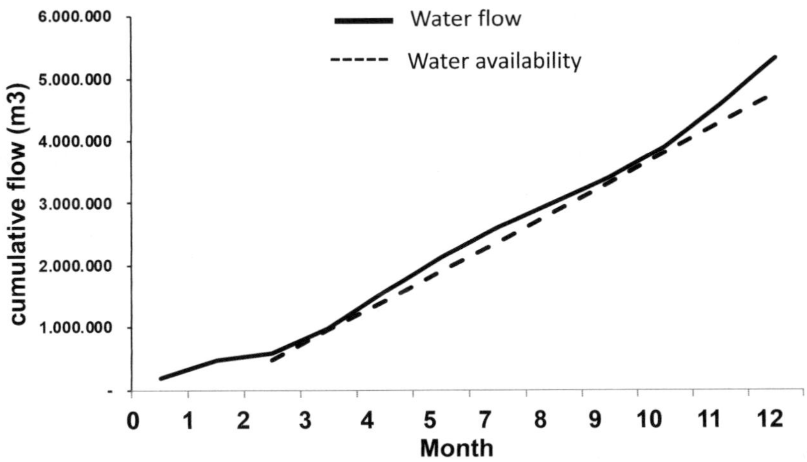

Figure 5. Water flow and water availability at Pulai reservoir catchment.

3.4. Alternative for capacity building for Pulai reservoir

Some alternatives were arranged in order to increase reservoir capacity to supply PDAM with raw water. The alternative was based on the calculation of reservoir water balance to determine the maximum amount of water that can be taken based on the incoming water, evaporation, and safe retrieval limits related to the intake and spillway positions. Alternative 1, was the intake at the current position and the safe limit on the volume of the 3,000,000 m^3 reservoir or reservoir water level of 1 meter above the intake. This alternative has been done. Alternative II, was the intake placed at 9 m or below 5 m from the current position with a safe limit volume of 2,500,000 m^3. Alternative III was the same as the second condition but the spillway position was raised by 1 meter.

The calculation results in alternative I showed that the change in storage has a negative value in January to April, and July to October 2009. It suggested that the direct inflow and rainfall were smaller than the discharge and evaporation from the surface of the reservoir. Assuming that the reservoir was fully filled in early January, then the lowest volume above the safe collection limit will occur in November. The maximum amount of water that can be taken was 590,000 m^3 per month (figure 6).

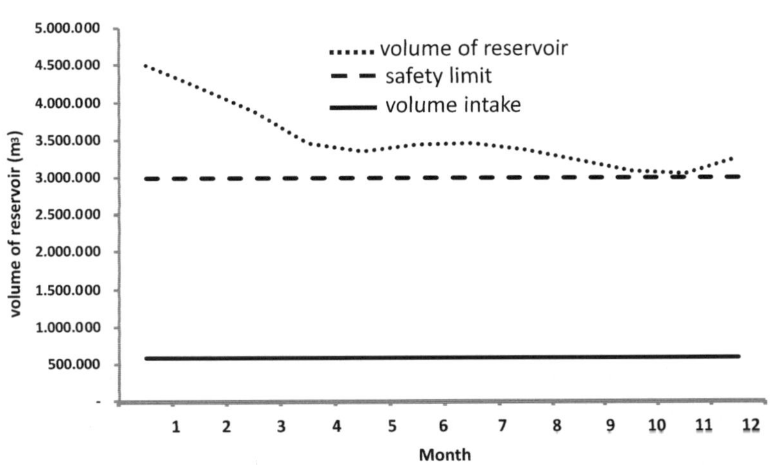

Figure 6. Volume of reservoir, safety limit and maximum volume intake on alternative I.

Alternative II showed that changes in storage have a negative value in January to April, and June to October 2009 (figure 7). This means that the direct inflow and rain fall were smaller than the intake discharge and evaporation from the surface of the reservoir. The maximum amount of water that can be taken was 640,000 m^3 per month. Assuming that the reservoir was fully filled in early January, then the lowest reservoir volume above the safe collection limit will occur in November. Safe volume is intended as a backup if there is a rain condition that is lower than normal conditions.

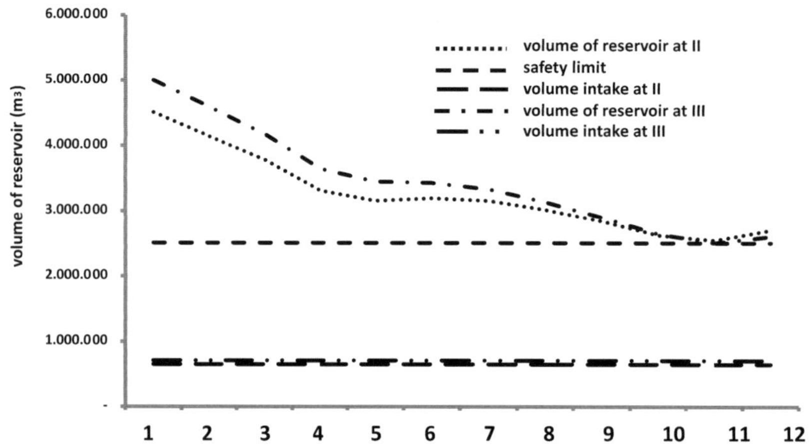

Figure 7. Volume of reservoir, safety limit and maximum volume intake on alternative II and III.

Alternative III showed that the maximum amount of water that can be taken to maintain the volume of the reservoir above the safe limit was 694,000 m^3. The amount of intake discharge resulted in negative storage changes from January to October 2009. This means that the direct inflow and rain fall were smaller than the discharge and evaporation from the surface of the reservoir. Assuming that the reservoir was fully filled in early January, then the lowest reservoir volume above the safe collection limit will occur in November (figure 7). In alternative III, the maximum amount of water that can be taken for PDAMs rose by 7.8% compared to alternative II and increased by 16.9% compared to the current condition (alternative 1).

4. Conclusion

Rainfall in 2009 (2,563 mm) which was far below the average of the last 29 years (3,216 mm) was the main factor that caused the water crisis and this condition was exacerbated by the conversion of 52% of the forest into shrubs and oil palm plantations in the Pulai reservoir watershed. A very large area of shrubs has the potential to be reforested, considering that this site as a protected forested area based on formal jurisdiction, so that it will increase water reserves. Meanwhile, the increase in air temperature above the average also contributed to the increase of water evaporation.

The current water supply capacity from the Pulai Reservoir to the PDAM is 590,000 m^3 per month with reservoir water reserves of 3 million m^3. If the intake position is lowered by 5 m and reservoir water reserves are 2.5 million m^3, the water extraction capacity will increase to 640,000 m^3/month and if added to increasing the spill way 1 m height can increase the water withdrawal capacity to 690,000 m^3/month.

IOP Conf. Series: Earth and Environmental Science **380** (2019) 012001 doi:10.1088/1755-1315/380/1/012001

5. References

[1] Rofi, Malbonis Salma and Munawar Cholil 2008 Analisis Pengaruh Perubahan Penggunaan Lahan Terhadap Debit Puncak Sub DAS Kali Premulung Tahun 2006 dan 2014. *Skripsi thesis, Universitas Muhammadiyah Surakarta*.Bab 1.

[2] Schwab Glenn, Fret Richard, Edminster Talcott, and Barnes K K 1981 Soil and water conservation engineering. New York, USA. Third edition p: 525

[3] U.S.Soil Conservation Service 1972 Hydrology. *National Engineering Handbook*. Section 4, Washington D.C. in : Sitanala Arsyad. 1989. Konservasi Tanah dan Air. *IPB*. Bogor

[4] Frich, Alexander, Delta-Marta, Gleason, Haylock, Klein Tank and Peterson 2002 Observed coherent changes in climatic extremes during the second half of the twentieth century. *Journal Climate Research*, Vol. 19: 193–212

[5] BMKG 2009 Data-data meteorologi stasiun meteorologi Tanjung Pinang - Bintan. *BMKG*

[6] Badan Lingkungan Hidup Kepri 2009 Kajian pemulihan lingkungan hidup catchment area sungai Pulai Provinsi Kepulauan Riau. *Laporan Akhir, Kerjasama Penelitian antara Pusat Penelitian Limnologi LIPI dengan BPL Provinsi Kepri*, Tanjung Pinang

[7] Seyhan Ersin 1979 Principles of reservoir engineering. *Institute of Earth Sciences*, Free University, Amsterdam, The Netherlands.

[8] Fakhrudin, M 2010 Evaluasi klimatologi krisis air di waduk Pulai – Bintan Kepulauan Riau. *Prosiding Seminar Nasional Penelitian, Pendidikan, Penerapan MIPA, Universitas Negeri Yogyakarta*, hal B - 204 – 210

[9] Fakhrudin,M. Iwan Ridwansyah, Fajar Setiawan, dan Meti Yulianti 2011 Evaluasi pola penggunaan lahan untuk konservasi waduk Pulai Bintan Kepulauan Riau. *Prosiding Seminar Nasional Hari lingkungan Hidup UNSOED dan Ikatan Ahli Lingkungan Hidup*, Purwokerto, Hal 150 – 156

[10] Soewarno, 2000. *Hidrologi Operasional*. Jilid I, Penerbit PT. Aditya Bakti, Bandung, Indonesia

[11] Starosolszky.O 1986 Basic phenomena of surface hydrology, in applied surface hydrology. *Water Resources Publications*. Colorado,USA. hal 3-30

[12] Arsyad, Sitanala 1989 Konservasi tanah dan air. *Penerbit IPB Press*. Bogor

IOP Conf. Series: Earth and Environmental Science **380** (2019) 012002 doi:10.1088/1755-1315/380/1/012002

Dissolved oxygen and temperature stratification analysis for early warning system development in preventing mass mortality of fish in lake Maninjau, West Sumatera - Indonesia

M. Fakhrudin[1*], Luki Subehi[1], Taofik Jasalesmana[1], Aan Dianto[1]

[1]Research Centre for Limnology, Indonesian Institute of Sciences, Cibinong 16911, Indonesia.

**Corresponding author :mfakhrudin@limnologi.lipi.go.id*

Abstract: Lake Maninjau is caldera lake which lies at Sumatera island, Indonesia. The objective of this paper is to analyse dissolved oxygen (DO) and water temperature (WT) stratification related to mass mortality fish. The WT and DO stratification were measured by logger version CTD profiler with a depth interval of 1 m in March 2018. Based on analysis at 12 stations of measurement from buoyancy frequency value showed various stratification. The top layer (epilimnion) had a thickness of 2 m and the concentration of DO was 4.8 mg/l – 13.5 mg/l. Meanwhile, the hypolimnion layer formed at various depths ranged from 9 - 20 m to the bottom of the lake with concentration of DO was close to zero. These two layers are separated by a thermocline (metalimnion), which acted as a buffer. When the surface temperature of the lake was cooler than the lower one, then the density was increased, the turnover occurred. Then, the concentration of DO at the upper layer of the lake decreased, caused mass mortality fish. The process of the mass water reversal was monitored through temperature stratification and DO measurements in real time, as criterion for early warning system of mass mortality of fish.

1. Introduction

Stratification of lake is a natural phenomenon and can change at any time, depending on the dynamics of local weather conditions. Changing stratification or stirring of water column is an ecosystem response to maintain flexibility and environmental capacity. In subtropical areas, the process of stirring lake water is more different clearly than that in the tropics. This stratification affects chemical and biological processes, such as nutrient dynamics, gas transfer and phytoplankton distribution in the water column [1, 2 , 3].

Hydrodynamics of lake water are influenced by lake morphometry, inflow and outflow, air factors, water density, and Coriolis effects [4]. In case of Lake Maninjau, the density of water and weather are the dominant factors, especially in the stratification dynamics. The strength of stratification is affected by the transfer of heat energy from sunlight to the body of water, air temperature and wind strength. If the sun's intensity is low and persist in a long time, the temperature of upper layer also decreases (the density of water increases), then the water is in the condition of the isotherm, thus facilitating the vertical diffusion process.

The seasonal variability of atmospheric conditions influenced fluctuations in air and water temperature nearly proportionally [5]. The results showed that increase of air temperature resulted in an increase of the epilimnion layer temperature [6] and changed the depth of the thermocline layer [7]. Stratification will disappear when solar thermal energy is not strong enough, then the kinetic energy of the wind can move warmer water from the upper to the lower layers [8]. There are several parameters used as indicators of the stability of lake water columns, including index stratification, buoyancy frequency, stability index, lake number, wedderburn number and Schmidt stability. Research on the stability of lake water columns has been widely carried out, especially in subtropical lakes. But a few study tropical lakes, especially in Indonesia. Recent study at Lake Maninjau, West Sumatra, Indonesia showed that there was a dynamic stratification of the daily water column, which was positively correlated with the power of sunlight and the strength of stratification of the water column was reduced when there was an increase in wind stress [9].

Lake Maninjau is Caldera lake which has a height of about 464 m above sea level. In addition, water surface area, coastline length, maximum depth and catchment area are 9,737 ha, 52.7 km, 168 m and 13,260 ha, respectively. Based on this morphometry, wind has an influence on the movement of water, especially on the surface and penetration of oxygen in water [10]. The relative depth (Zr) is 1.508 and the Shore line development (DL) is 1.51 km/km^2 [11]. The Zr value means low stability, easy to experience stirring due to outside influences, such as strong winds. Lake has a high stability value of Zr > 4, and is a deep lake with a narrow surface area [12]. Meanwhile, the DL value shows that the waters of the lakeside areas do not support the productivity of their waters. The longer the coastline the greater the productivity [13].

At present, Lake Maninjau, not only for a power plant that generates annual energy of 205 GWH, but also used for floating net cages (KJA) which number more than 15 thousand plots. Every year, there are frequent mass mortality of fish that coincide with upwelling process. Research Centre for Limnology LIPI in collaboration with relevant agencies has been establishing online environmental parameter monitoring systems and developed into an early warning system for mass mortality fish. Therefore, this paper aims to examine the dynamics of temperature stratification and dissolved oxygen (DO) in Lake Maninjau for the basis of further development of early warning systems.

2. Materials and methods

The logger version CTD profiler with optical fast DO sensor RINKO-Profiler was used for our survey. Depth (semiconductor pressure sensor with ranged 0 to 600 m and resolution 0.01 m), temperature (thermistor with ranged -3 to 45°C and resolution 0.001°C) and dissolved oxygen/DO (phosphorescence with ranged 0 to 20 mg/L and resolution 0.001mg/L) were obtained at 61 stations. Measurements were carried out on March 12 – 17, 2018 with an interval depth of 1 m. The measurement station is presented in figure 1. Meanwhile, the data series of temperature profile and DO period starts from January to May 2018 obtained from e-most (monitoring on line) of Research Centre for Limnology station.

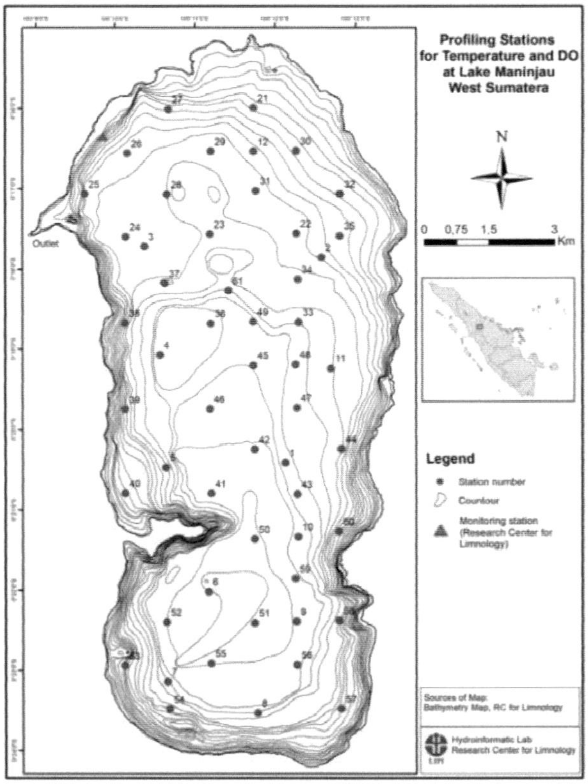

Figure 1. Stations of temperature stratification and DO at Lake Maninjau.

2.1. Determination of the stability of the water column

a. Buoyancy Frequency

Buoyancy frequency (N), is the frequency of oscillation of a water package in a stable layer of water column. This size is an indicator of water stability. The greater the frequency, the more stable of the water column. Buoyancy frequency, which is the local stability calculated by referring to [3].

$$N = \sqrt{\frac{g}{\rho}\frac{\partial \rho}{\partial z}}$$

where

ρ : (kg/m^3) Water density for each depth layer (kg/m^3)

$\frac{\partial \rho}{\partial z}$: gradient density ((kg/m^4)

g : gravity acceleration (m/s^2)

b. Stratification Index (SI)

The stratification index is the potential energy needed to mix the water column. If the water column has a large stratification index, then the large effort needed to mix the water column (water column is stable). Conversely, if the water column has a low SI, it will be easy to mix the water column. Next, the calculated index stratification refers to [14].

$$SI = -\int_{-h}^{0} \left(\rho - \langle \rho \rangle\right) gz dz$$

where

ρ	: the density of water in each layer of depth (kg/m^3)
$<\rho>$	: average density of water for each layer of depth (kg/m^3)
g	: gravity acceleration (9,8 m/s^2)
$(T_0 - T)$	: delta of water surface and air temperatures

3. Results and discussion

3.1. Stratification and distribution of water temperature

In general, the interaction process of physics, chemistry and biology in water is closely related to the function of temperature. Temperature is also an important parameter in determining the level of habitat feasibility for fish and other organisms. Water temperature plays an important role in the distribution of species [15], and the growth and survival of aquatic organisms, including phytoplankton and fish [16, 17 , 18]. Water temperature also affects the reproductive time, development, growth, death and metabolism of most species, including fish [19, 20, 21 , 22].

Moreover, water temperature determines the density of water (ρ), while ρ is the basic parameter in hydrodynamics. The difference ρ results in stratification of the water layer (epilimnion-metalimnion-hypolimnion) and prevents vertical mixing. According to [23], ρ is determined significantly by temperature, salinity, and total concentration of suspended sediments. However, according to [24], suspended sedimentary factors are important in ice lakes, while salinity factors only play an important role in high salinity lakes.

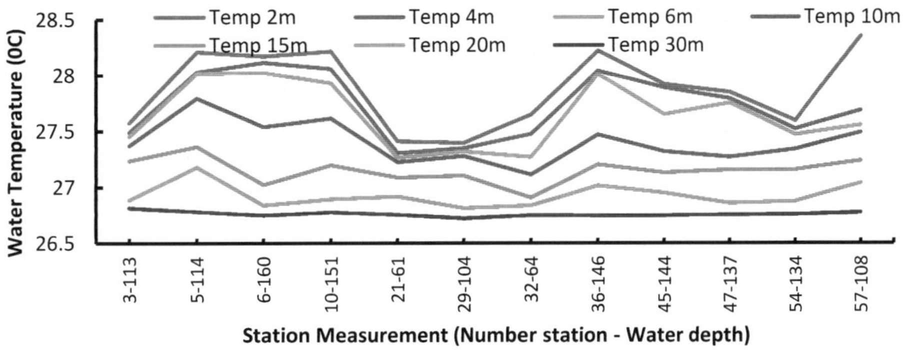

Figure 2. Water temperature profiling at Lake Maninjau.

For more detailed analysis, 12 stations were chosen from 61 stations which could represent both the aspect of space (horizontal distribution) and time (measurement in hours) and depth of the lake. The measurement results are shown in figure 2.

Figure 2 shows the water layer to a depth of 6 m, had a horizontal distribution of differences more than 1°C. Similarly, that at a depth of 10 m, 15 m and 20 m were 0.6°C, 0.3°C and 0.2°C, respectively. The smallest difference at 30 m depth, seemed evenly distributed throughout the lake. The temperature stratification at the 12 stations also showed the same pattern to a depth of 20 m. In addition, the fluctuation of water temperature decreased with increasing depth, indicating homogeneous and constancy of temperature in the deepest layer (table 1).

IOP Conf. Series: Earth and Environmental Science **380** (2019) 012002 doi:10.1088/1755-1315/380/1/012002

Table 1. Average and standard deviation of temperature (Temp) for each depth.

Temp (°C) – Depth (m)	Standard deviation (σ)
Temp 2 m	27.9 °C ± 0.348
Temp 4 m	27.7 °C ± 0.296
Temp 6 m	27.6 °C ± 0.298
Temp 10 m	27.4 °C ± 0.188
Temp 15 m	27.2 °C ± 0.115
Temp 20 m	26.9 °C ± 0.106
Temp 30 m	26.8 °C ± 0.024

Stratification of water layer based on temperature shows less clear boundaries, especially between the epilimnion and metalimnion, however the stratification for the hypolimnion layer was clearer (figure 3). This unclear stratification is common in tropical regions, so it is difficult to identify the epilimnion layer [25]. At a depth of 20 m, the water temperature of 26.8 °C showed insignificant differences, so this layer acted as a hypolimnion. The depth water of 1.5 m was very dynamic. Compared with a depth of 2 m, considered to be more stable, so the temperature difference with the hypolimnion layer also varies greatly, ranging from 0.6 °C (Stations 21-61 and Stations 29-104) - 1.6 °C (Stations 57-108) (figure 2).

In the hypolimnion depth, the density of water showed a value of 0.997 gr/cm3. The values of water density compared with water surface showed small values, ranging from 0.0003 - 0.0023 gr / cm3 (Station 3-113 ρ 0.9964 gr / cm3 and Station 47-137 ρ 0.9964 gr / cm3) (figure 4). The difference of ρ between the hypolimnion layer and the small above layer indicated no stirring of water mass. According to [24] the density with difference of 0.0003 is sufficient to prevent mixing. So, the measurement on March, the area of Station 47-137 has a very strong water period stratification compared to the Station area 3-113. This condition indicated uneven stirring occurred in whole lake area.

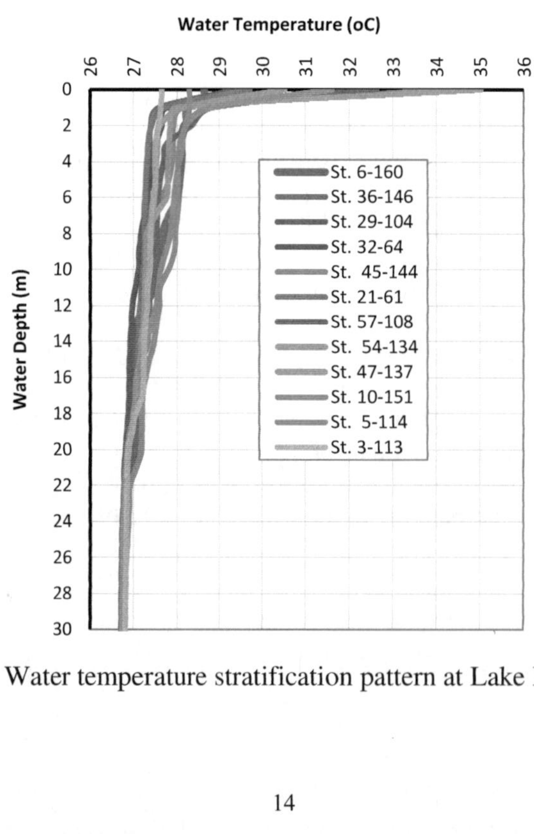

Figure 3. Water temperature stratification pattern at Lake Maninjau.

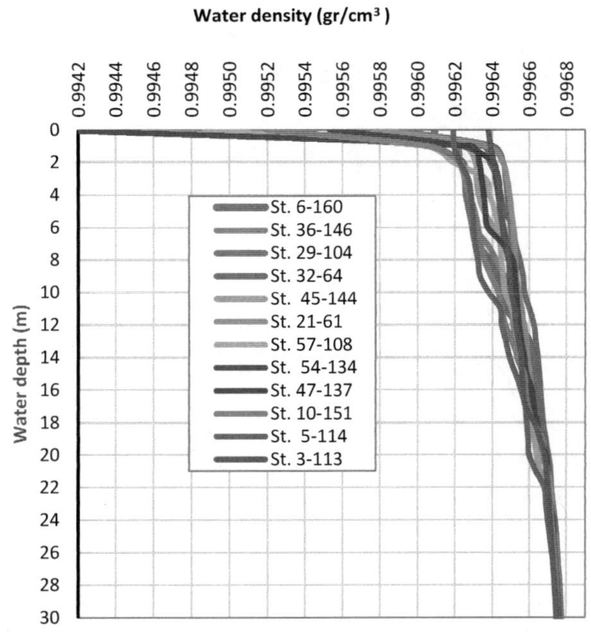

Figure 4. Water density stratification pattern at Lake Maninjau.

3.2. Stratification and distribution of DO

Figure 5 shows the variation of DO values on the lake surface had a large value and decreased followed the depth. At 10 m depth, the DO values have started to zero at some stations, and starting at a depth of 15 m, only a few stations whose values are close to 2 ppm (figure 5). At Station 21-61 DO at a depth of 2 meters around 3.8 ppm, and a depth of 4 m has decreased to 2.5 ppm. This area is KJA region, which requires quite a lot of oxygen, resulting in a decrease in DO. In the case of low DO content, it has caused disturbances in fish that DO conditions above 5 ppm for the growth of all aquatic organisms [26].

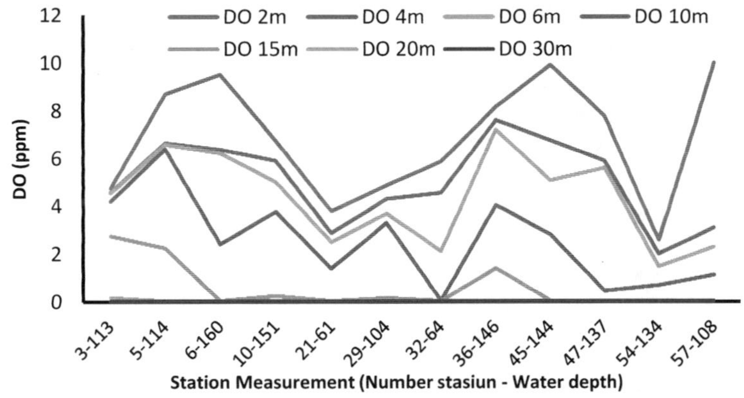

Figure 5. DO profiling at Lake Maninjau.

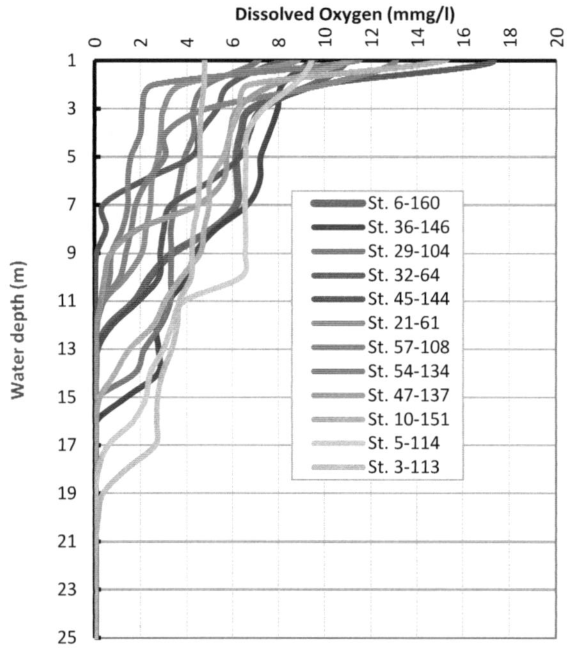

Figure 6. DO stratification pattern at Lake Maninjau.

DO stratification profiles have the same pattern as water temperature (figure 6). Based on 12 stations, it showed DO zero varies in water depth from 7-19 m to the bottom. At a depth of 2 m several stations have shown low DO values, such as Station 21-61 DO 3.7 ppm and even at Station 54-134 DO 2.6 ppm. It seemed that Lake Maninjau at a water depth of up to 19 m had a dynamic DO variation, while the water depth of more than 19 m DO have been zero throughout the lake.

3.3. Water column stability

Figure 7.A showed the highest buoyancy frequency (St.45-144) and the lowest (St.57-108) at Lake Maninjau. The first maximum peak of the N curve at both stations was at a depth of 5 m and 15 m. The peaks of the curve showed that at this depth, there is a thermocline layer which can prevent the mixing and diffusion process. There was a peak variation of the N curve from the surface to 30 m. From a depth of 30 m to the bottom, the frequency was the same at each depth. It indicated the density of water in each layer of water was homogeneous. There was no density gradient indicating that there was no barrier to mixing and diffusion processes. It seemed the depth range of 20 m – 30 m, it was started for homogeneous layer until bottom of lake.

In addition to buoyancy frequency, the stratification strength of Lake Maninjau was also characterized by using the stratification index (figure 7.B). The calculation results showed that the sampling point of the measuring water column had SI ranging between 14.47 - 4103.86 Jm-2. The greatest strength of stratification was in the middle (St. 45-144 and St.47-137), while the smallest in the south. This corresponds to the N curve, which was at St. 45-144 N curve had the highest peak compared to other stations. The high SI value at this point was caused by the amount of heat energy absorbed in the area and this condition corresponds to the high water temperature in the area. This situation causes the penetration of heat energy deeper and shifts the thermocline layer deeper. As a result, water stability will increase.

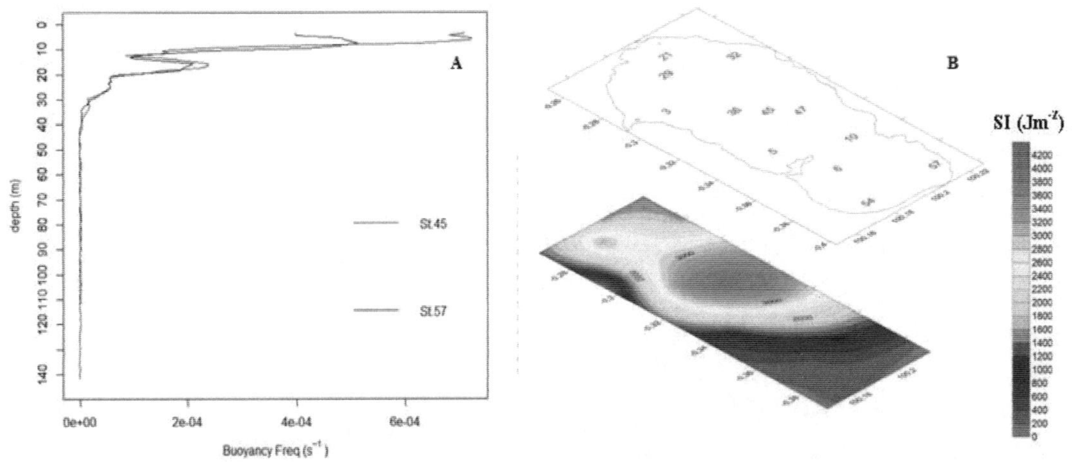

Figure 7. A. Buoyancy frequency at Station 45-144 and Station 57-108. B. Distribution of stratification index at Lake Maninjau.

3.4. Mass mortality fish phenomena and early warning system

Based on the results above, there were variations in temperature and DO profiles both vertically and horizontally that vary from one location to another. The stirring process also varies, not simultaneously throughout the lake. This also shows that when the stirring occurred, only fish in the fish cage died, because the fish trapped in and could not look for locations that did not occur stirring/oxygen in the other places.

Figures 8 and 9 show the beginning of the year, the water temperature started at a depth of 4 m almost homogenous. This condition occurs most often in January around 12 days and decreased from February to May. At the same time, DO also dropped, it was close to zero and caused fish died, especially on fish cage. This stirring of water was also triggered by the cold air temperature in a few days earlier. In January, the difference in daily air temperature with the water surface ranged from 0.2 to 0.4 °C. Whereas in other conditions the daily temperature difference reaches 1.8 °C.

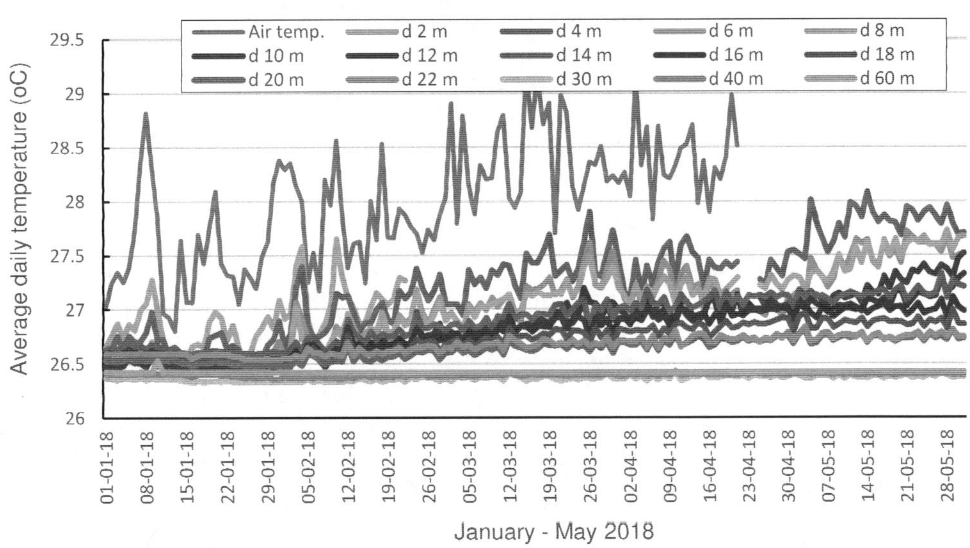

Figure 8. Daily temperature stratification pattern at Lake Maninjau (January-May 2018).

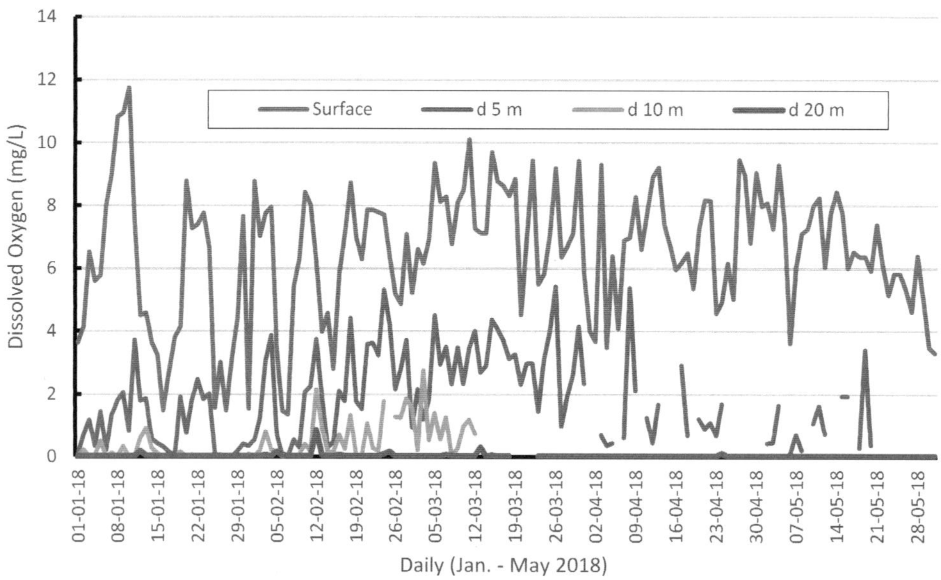

Figure 9. Daily DO stratification pattern at Lake Maninjau (January-May 2018).

Figure 10. Early warning system for mass mortality fish monitoring.

Figure 10 showed an early warning system that established by the Research Centre for Limnology LIPI at Lake Maninjau-West Sumatra. Principally, this system records the water temperature and oxygen from surface to a depth of 60 m with interval measurement 10 minutes. The measurement results in the field are immediately sent in real time to the server at the Cibinong-Bogor office. Data entered in server is processed and compared to certain criteria whether it is categorized as hazard, standby 1, standby and normal. Each change in status (hazard, standby 1, standby and normal) always gives a signal to the registered cell phone. The results of temperature monitoring and DO data can also be accessed through website [27].

4. Conclusion

Water temperature and DO stratification in Maninjau Lake are very dynamic, especially up to a depth of 6 m, which was characterized by horizontal (space) and vertical (depth) variations that had wide those vulnerable values. The hypolimnion layer started at a depth of about 20 m - 30 m which was evenly distributed throughout the lake, while the upper layer was difficult to separate between the epilimnion and metalimnion. By understanding the dynamics of temperature stratification and DO in detail, it can be used as a reference for the development of warning systems.

5. References

[1] Yang Y, Colom W, Pierson D, Pettersson K 2016 Water column stability and summer phytoplankton dynamics in a temperate lake (Lake Erken, Sweden). *Inland Waters* 6: 499–508. DOI: 10.5268/IW-6.4.874

[2] Ladd C, Stabeno PJ 2012a Deep-Sea Research II Stratification on the Eastern Bering Sea shelf revisited. 70: 72–83. DOI: 10.1016/j.dsr2.2012.02.009

[3] MacIntyre S, Sickman JO, Goldthwait S a., Kling GW 2006 Physical pathways of nutrient supply in a small, ultraoligotrophic arctic lake during summer stratification. *Limnology and Oceanography* 51: 1107–1124. DOI: 10.4319/lo.2006.51.2.1107

[4] Tsanis, Ioannis K, Jian Wu, Huihua Shen 2007 Hydrodynamic and pollutant transport modeling of lakes and coastal waters. *Environmental hydraulics: Elsevier*

[5] Subehi L, Fukushima T, Onda Y, Mizugaki S, Gomi T, Terajima T, Kosugi K, Hiramatsu S, Kitahara H, Kuraji K, Ozaki N (2009) Influences of forested watersheds conditions on fluctuations in stream water temperature with special reference to watershed area and forest type. *Limnology* 10: 33 – 45

[6] Magee, Madeline R., and Chin H. Wu 2016 Response of water temperatures and stratification to changing climatein three lakes with different morphometry. *Hydrol. Earth Syst. Sci.* Discuss., doi:10.5194/hess-2016-262

[7] Schindler, D. W., Beaty, K. G., Fee, E. J., Cruikshank, D. R., Debruyn, E. R., Findlay, D. L., Linsey, G. A., Shearer, J. A.,Stainton, M. P. and Turner, M. A 1990 Effects of climatic warming on lakes of the central boreal forest. *Science*,250(4983), 967–970, doi:10.1126/science.250.4983.967, 1990

[8] Tuan , N. V., K. Hamagami, K. Mori Y. Hirai 2009 Mixing by wind-induced flow and thermal convectionin a small, shallow and stratified lake. *Paddy Water Environ* 7:83–93

[9] Jasalesmana,Taofik; Zulti, Fifi; Suryono,Tri ; Santoso AB 2017 The seasonal of thermal stratification and water column stability of lake Maninjau, West Sumatera. *Proceedings of the 16th World Lake Conference:* 593–599

[10] Fakhrudin, M., H. Wibowo, L. Subehi, dan I. Ridwansyah 2002 Karakterisasi hidrologi danau maninjau Sumatera Barat. *Prosiding Seminar Nasional Limnologi*. Pusat Penelitian Limnologi – LIPI

[11] Fakhrudin, M 2011 Identifikasi hidrodinamika sebagai dasar pengembangan sistem peringatan dini bencana kematian massal ikan di danau Maninjau Sumatera Barat. *Prosiding Seminar Nasional Penelitian, Pendidikan dan Penerapan MIPA, Universitas Negeri Yogyakarta*

[12] Wetzel, R. G 1983 Limnology. *W. B. Saunders College Publ.,* Philadelphia

[13] Welch, P. S 1952 Limnology. *Mc Graw-Hill Book Company, Inc*

[14] Ladd C, Stabeno PJ 2012b Stratification on the Eastern Bering Sea shelf revisited. *Deep Sea Research Part II: Topical Studies in Oceanography* 65–70: 72–83. DOI: 10.1016/j.dsr2.2012.02.009

[15] Magnuson JJ, Crowder LB, Medvick PA 1979 Temperature as an ecological resource. *American Zoologist* 19: 331 – 343

[16] Christie GC, and Regier HA 1988 Measures of optimal thermal habitat and their

relationship to yields for 4 commercial fish species. *Canadian Journal of Fisheries and Aquatic Sciences* 45: 301 - 314

[17] Magnuson JJ, Meisner JD, Hill DK 1990 Potential changes in the thermal of Great Lakes fish after global climate warming. *Transactions of The American Fisheries Society* 119: 254 - 264

[18] Jackson DA, Peres-Neto PR, Olden JD 2001 What controls who is where in freshwater fish communities – the roles of biotic, abiotic, and spatial factors. *Canadian Journal of Fisheries and Aquatic Sciences* 58: 157 - 170

[19] Shuter B.J, Post JR 1990 Climate, population viability, and the zoogeography of temperate fishes. *Transactions of The American Fisheries Society* 119: 314 – 336

[20] Tonn WM 1990 Climate change and fish communities: a conceptual framework. *Transactions of The American Fisheries Society* 119: 337 – 352

[21] Brandt SB, Mason DM, McCormick MJ, Lofgren B, Hunter TS 2002 Climate change: implications for fish growth performance in the Great Lakes. *American Fisheries Society Symposium* 32: 61 – 76

[22] Casselman JM 2002 Effects of temperature, global extremes, and climate change on year-class production of warmwater, coolwater, and coldwater fishes in the Great Lakes basin. *Fisheries in a Changing Climate (Ed. N.A. McGinn)*, pp. 39 – 6O. Blackwell Publishing, Bethesda, Maryland

[23] Ji, Zhen-Gang 2008 Hydrodynamics and water quality: modeling rivers, lakes and estuaries. *Wiley-Interscience*

[24] Reading, H. G 1996 Sedimentary environments: processes, facies, and stratigraphy. *Wiley-Blackwell*

[25] Lewis JR, William M 1987 Tropical limnology. *Annual Review of Ecological Systems* 18: 159 - 184

[26] Effendi, H 2003 Telaah kualitas air bagi pengelolaan sumber daya dan lingkungan perairan. *Kanisius*, Yogyakarta, 249 hlm

[27] Fakhrudin, M., Ika A.S, Tjandra C, A. Hamid, Foni A.S, Endra T, dan Kodarsyah 2012 Pengembangan sistem monitoring online dan peringatan dini bencana lingkungan: studi kasus di danau Maninjau. *Prosiding Seminar Nasional Limnologi*. Pusat Penelitian Limnologi – LIPI

IOP Conf. Series: Earth and Environmental Science **380** (2019) 012003 doi:10.1088/1755-1315/380/1/012003

Application of floating treatment wetlands in a highly eutrophic lake, Indonesia: a new tool for lake restoration and provision of microhabitat

C Henny[1*], R Kurniawan[1] and Trisuryono[1]

[1]Research Centre for Limnology, Indonesian Institute of Sciences, Jl. Raya Bogor km 46, Cibinong, Indonesia

**Corresponding author: cynthia@limnologi.lipi.go.id*

Abstract. Two sets of floating treatment wetlands (FTWs) have been evaluated as a tool for lake restoration and microhabitat provision for aquatic biota in a highly eutrophic and polymictic Lake Maninjau, Indonesia. Each FTW system consisted of 14 units square floating frame built using PVC pipes and palm sugar fibres as material mat transplanted onto net. All units of FTW were planted with different vegetation composition arrangement then placed inside the square metal frame attached on buoyant plastic drums to support the FTWs against the wave. The plants used in this study were *Canna variegata*, *Vetiveria zizanioides*, *Cyperus papyrus*, *Echinodorus polaefolius* and *Limnocharis flava*. FTWs were evaluated for 7 months for nutrient removal efficiency and other water quality indicators. Despite unstable lake conditions accompanied by strong waves and lake mixing, the cage and the PVC frame of FTWs applied remained intact and providing the buoyancy to support the plant growth. The plants in FTWs grew and adapted well in the turbulence water and even in the hypoxic conditions. Improvement of lake water quality in the FTWs area may be small but the accumulated nutrients content in the plant shoots suggests high removal of nutrients. In total by the end of observation, one FTWs system could remove > 100 g N and > 7 g P from the lake water. During field observation, important local fish and diverse wildlife activities were recorded including inhabiting, foraging, breeding, nursing and resting in the FTWs. Our study suggests that FTWs can be a promising tool for lake restoration to improve lake water quality and for microhabitat provision.

1. Introduction

Lake eutrophication has been a worldwide problem due to excessive input of nutrients (nitrogen and phosphorus) derived from anthropogenic sources [1,2,3,4]. One of the major causes of lake eutrophication faced by Indonesian lakes is intensive floating cage aquaculture [3,5,6]. Excess nutrient enters the lake will likely to be accumulated in the sediment especially phosphorus which subsequently will be released back to the water column via microbial transformation [7]. Both external and internal loadings elevate P content in the lake triggering the harmful blue-green algal blooms. Excessive algal growth reduces water quality mostly dissolved oxygen (DO) level. Depletion of DO can cause fish kill and loss of other aquatic life [8,9]. Although it is not environmentally sound, the cage aquaculture has been a big challenge to be banned due to an economic issue for the local people. Undertaking effort to improve water quality by reducing the activity of cage aquaculture has been very slow. Therefore, an innovative treatment tool is highly needed not only to reduce excess nutrient directly in the lake water but to provide the microhabitat as well to conserve aquatic life especially the economic local fish.

Floating treatment wetlands (FTWs) are a new innovative tool, a novel treatment concept that can be suitable to be applied to restore the lake suffering from eutrophication. FTWs offer a promising technique, low cost green approach for lake water quality improvement and microhabitat provision as well as aesthetic improvement at ponds or small lakes [10,11,12,13]. FTWs are basically cultivated emergent plants growing on a buoyant mat which can be deployed in the water bodies. The plant roots will develop down in the water column, creating root system which is important for the performance of the FTWs on the nutrient uptake and pollutant transformation. The root network system not only functions as a filter to entrap particulates but also provide extensive surface area for biofilm which is responsible for the biochemical transformation of contaminants [13,14,15, 16]. Several studies have demonstrated the effectiveness of FTWs on nutrient removal or other type of contaminants from various wastewater sources. The studies were done mostly in mesocosm laboratory scale [14,15] and in field application in ponds [16,17,18], but limited field studies were conducted in the river [11]. To the best of our knowledge no study has been reported in the big lake. The study mostly used a single type of plant, but limited study used a combination of two or several types of plants [12,14,19].

The application of FTWs for big lake restoration will face substantial challenges due to lake hydrodynamics. The lake is very susceptible to the influence of climate variability. Frequent lake mixing due to strong wind and big wave will affect the sustainability of FTW system. The phenomenon of lake upwelling triggering anoxic and rich nutrient bottom water pushed upward to the surface may cause a shock loading for the root system of FTWs. We investigated the FTW system in a highly eutrophic and polymictic Lake Maninjau, Indonesia for lake restoration and microhabitat provision as well. Lake is very unstable and susceptible to mixing due to the turbulence weather. We used a combination of four different type of emergent plants. The lake has been in hyper-eutrophic conditions for over a decade due to overloading of nutrient and organic matter from the waste input of intensive floating cage aquaculture. The lake water conditions were worsened by dead fish mass due to frequent fish kill incidents, and harmful algal blooms. This causes continuous decline on important local fish with economic value [3,7,8]. The objectives of this study were to (1) evaluate the performance of the FTW system on lake water quality improvement (2) investigate the sustainability of FTW system and the selection of plants (3) observe wildlife utilizing the FTW system for habitation or feeding.

2. Materials and methods

2.1. Study site

The FTWs study was conducted at Lake Maninjau, a caldera lake located at Agam District in the Western area of Sumatra, Indonesia (S 0°22'33.0", E 100°11'35.1"). The lake is considered a big lake which has surface area of 9737.5 Ha and maximum depth of 165 m. The lake is utilized for hydropower and intensively for floating cage aquaculture. There were more than 17000 units of floating cage aquaculture reported throughout the lake periphery [7]. The lake is highly eutrophic with elevated nutrient concentration and frequent mass fish kill incidence reported several times in a year generally in rainy (November – February) and transitional season (March – May) during high climate variability. High climate variability induces strong wind with heavy rain causing lake mixing and lake upwelling subsequently. Figure 1 shows the map of Lake Maninjau and the location where the FTWs were set and the pictures of indication of heavily eutrophic conditions of lake water with dead fish. One FTW was placed at the periphery of the lake outlet area of Muko-Muko (FTW_MK) where it is close to the hydropower turbine in which the water current is relatively stronger. The other FTW was placed at Bayur area (FTW_BYR) where the fish cage units were very dense and less affected by the water current.

2.2. Design of Floating Treatment Wetland (FTW) systems

Two FTWs systems were applied in Lake Maninjau in August 2016. The system each consisted of 14 units of small FTWs. FTWs system was planted with different vegetation composition arrangement then was placed inside the square metal frame attached on buoyant plastic drums to support the FTWs against the wave (figure 2). The net (mesh 2 x 2.5 cm) was put underneath as a fence to protect the plant roots grazing. Each unit of FTWs (1m x 0.6 m, L x W) built with PVC pipes (3-inch Ø) (square floating frame) as buoyant material, palm sugar fibres as material mat transplanted onto net (mesh size of 4 x 4 cm).

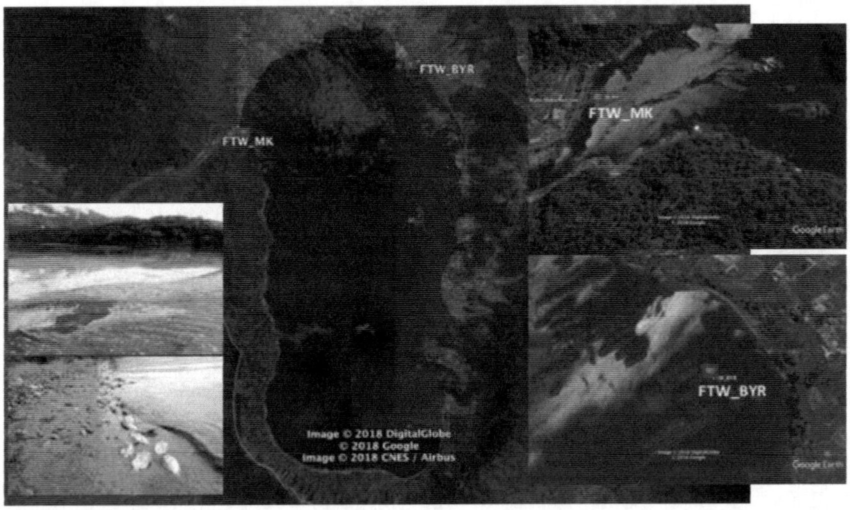

Figure 1. Map of Lake Maninjau (middle) and the FTWs sites (Muko-Muko area (FTW_MK), top right and Bayur area (FTW_BYR), bottom right)). Indication of lake eutrophication with thick algal bloom and dead fish (left) (right) (Google Earth, 2018).

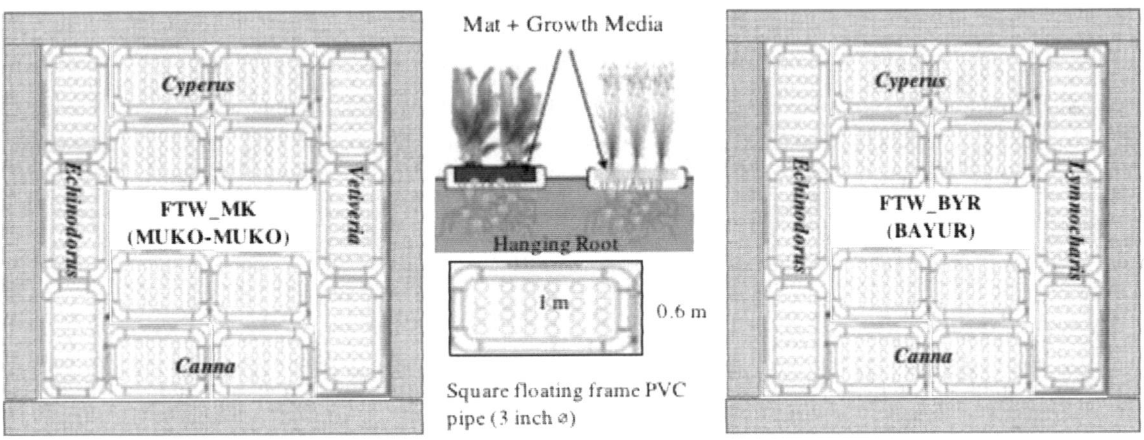

Figure 2. Schematic of Floating Treatment Wetlands (FTWs) system with different vegetation composition for each site. Plan-view (left and right) and cross-section of FTW unit (middle).

Each unit FTW has 15 plants with single plant grown in the individual hydroponic plastic cup (6 cm Ø (bottom) x 15 cm (height) x 10 cm Ø (top)) filled with growth media. The plastic cup was removed from the plant before being planted on fibres mat. The compost was added on the mat around the plant. The plants used in this study were *Canna variegata* (4 units; total of 60 plants), *Vetiveria zizanioides* (4 units, total of 45 plants), *Cyperus papyrus* (4 units; total of 60 plants), *Echinodorus palaefolius* (small type) (4 units, total of 45 plants) for the FTW_MK and *Canna variegata* (4 units; total of 60 plants),

Limnocharis flava (4 units, total of 45 plants), *Cyperus papyrus* (4 units; total of 60 plants) and *Echinodorus palaefolius* (4 units, total of 45 plants) for the FTW_BYR.

2.3. Sampling and lake water analysis
Direct water quality measurement and water sampling were conducted one on the inside of FTW system and one on the lake not far from the FTWs (over 10 m distance) one month after the FTWs was well established which was started in September 2016, in October 2016 then in January, February and March 2017. The water was collected at a depth of at least 20 cm. Physicochemical parameters were: pH and conductivity, measured directly in the water column using a Water Quality Meter (Horiba U-10), while temperature and dissolved oxygen (DO) were measured using a DO meter (YSI 6000). The evaluated parameters were: Total Nitrogen (TN), Total Phosphorus (TP), and Total Kjeldahl Nitrogen (TKN), analysed using spectrophotometric methods; chlorophyll-a, prior to lab analyses water sample was added with saturated $MgCO_3$ and filtered by using Whatman (GF/F) filter paper. Filtrates were then extracted and measured using a standard spectrophotometric technique; total suspended solids (TSS) determined using a gravimetric method. All lab analyses procedures were done according to APHA standard methods [20]. The samples prepared by spectrophotometric methods were measured by using Shimadzu UV 2100 Spectrophotometer.

2.4. Plant growth and microhabitat provision
The growth of plants was observed by measuring randomly shoot length (*C. variegata*), the shoot length and number (*V. zizanioides* and *C. papyrus*), the shoot length (*E. palaefolius* and *L. flava*). The root length was measured on September 16 and in March 2017 and plant biomass (dry weight) of the plants were estimated from samples data in January 2017 and in March 2017. Biomass estimation (dry weight analysis) was conducted only on the above parts of plants (shoot) for *C. variegata*, *V. zizanioides* and *C. papyrus* where 3 selected plants were harvested for each type of plants and air dried and then oven dried (50°C) until the constant weight was obtained. Plant productivity was estimated according to Olguìn [17].

2.5. Statistical analysis
The mean and standard deviation were calculated for the plant shoot length and number and root length. Statistical analysis of a simple t-test was done to compare the plant growth of two FTW systems studied. One-way ANOVA was also conducted to compare water quality data between FTW systems and the lake water in a different location. Pearson correlation analyses were also conducted for TN, TP and chlorophyll-a data. All statistical analyses were conducted with SPSS (24 version) and Microsoft Office Excel (2013).

3. Results and discussion

3.1. Water Quality
Physicochemical data are shown in figure 3. Water temperatures in the FTWs ranged from 27 to 29 °C which were slightly cooler than the lake water temperatures (28 to 30 °C). The water temperatures are much cooler in October 2016 and February 2017. September through February are usually the period of the rainy season and March is the period of transition season from the wet to the hot seasons in the lake area [8]. The pH values ranged from 6.8 to 8.2; conductivity ranged from 125 to 160 μS/cm while dissolved oxygen (DO) ranged from very low of 1 mg/L to the highest of 8 mg/L in the lake water and in the FTWs systems. All the physicochemical parameters measured were not shown significant difference between the lake water and the water in the FTW systems, although DO concentrations in the FTWs and in the lake water at the outlet area (MK) were much higher than those in the FTWs and in the lake at the area with dense floating fish cages (BYR). Very low DO concentrations of 1- 2 mg/L were recorded in September, October 2016 and February 2017 indicating the lake was in hypoxic condition [1,3,6]. At the time of measurements in those months, it was also recorded the upwelling phenomena

occurrence which caused the mass fish kill. The lake appeared to have needed longer time to recover from this hypoxic condition. In fact, DO level could only reach as high as 4 mg/L until March 2017. Slightly elevated DO of 8 mg/L was recorded at the outlet lake area which probably was due to stronger water flow. The hypoxic condition severely impacts the lake conditions by creating a stressful environment for aquatic life especially local fish, shrimps and molluscs. The growth of the plants in the FTWs and associated organisms such as microbial biofilm will also be affected by these hypoxic condition [18]. During this hypoxic condition the local fish, shrimps and molluscs were observed on the water surface and attached on the FTWs mats suggesting the presence of FTWs could support aquatic life during the lake extreme condition.

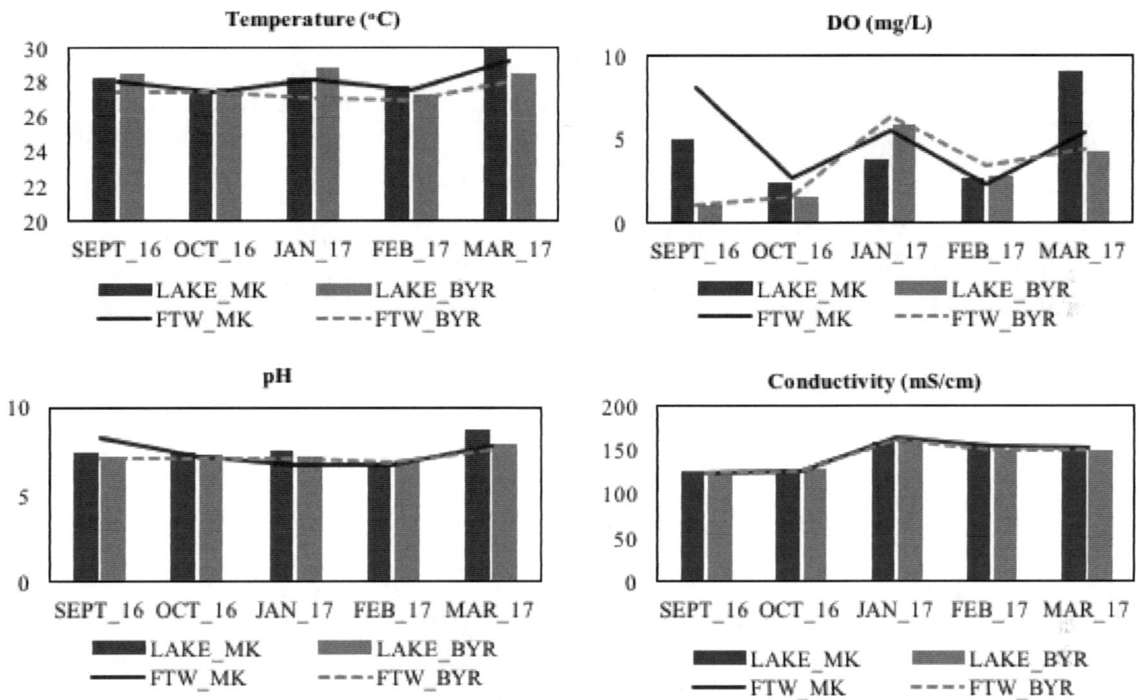

Figure 3. Temporal variation of temperature, pH, dissolved oxygen (DO) and conductivity in the lake water and the water inside the FTW system at two locations.

Temporal variation of concentrations of total nitrogen (TN), total phosphorus (TP), chlorophyll-a and total suspended solids (TSS) in the lake water and FTW systems are shown in figure 4. The results of the ANOVA test showed no significant different on nutrients, chlorophyll-a and TSS concentrations between the lake water and the water in the FTW systems at two locations ($p>0.05$) although some numbers for TN, TP and chlorophyll-a were relatively higher between the lake water and between the FTWs in two locations. Calculated removal efficiency (RE) based on the value difference of the FTWs and the lake water for each location showed high TN removal of around 82 % and 40% in September 2016 and in March 2017 respectively for the FTW_BYR while for the FTW_MK, TN removal reached only as high as 25 %. TP removal efficiency reached as high as 54% for the FTW_MK in October 2016 and in March 2017 while the FTW_BYR only had < 15 % of TP removal. Both FTWs gave around 50 % of chlorophyll-a removal in October 2016 and in March 2017. Both of FTWs did not have good removal efficiency (RE) for TSS. This probably is due to the lake often suffers from extreme weather causing waves and turbulence in the waters. Our study gave close removal percentage of TN and TP with the FTWs of the previous study [14]. The FTW field study using a combination of *Cyperus papyrus* and *Pontederia sagittata* in a eutrophic pond gave phosphate and nitrate removal percentage with the highest average of 76 % and 67 % respectively [19]. TN, TP and chlorophyll-a concentrations were around < 1.5 mg/L, < 0.15 mg/L and < 25 mg/L respectively in the FTWs and in the lake waters in this study. These numbers were similar to the numbers reported in Lake Maninjau and eutrophic ponds of previous studies [3,8,17,18, 19]. The chlorophyll-a concentrations measured in this study were considered low for Lake Maninjau compared to the previous studies results. Chlorophyll-a concentrations in Lake Maninjau could reach as high as 75 - 100 mg/L during hypereutrophic condition [3,8]. More cloud periods and hypoxic condition at the time of observations influenced the phytoplankton productivity and slowed the photosynthetic process [19]. Chlorophyll-a was positively correlated with TP with a high Pearson correlation coefficient of 0.627 ($p<0.05$) and had a weak correlation with TN (table 1). The results indicate that phosphorus was the limiting factor for phytoplankton productivity and controlled the chlorophyll-a concentration [3,21].

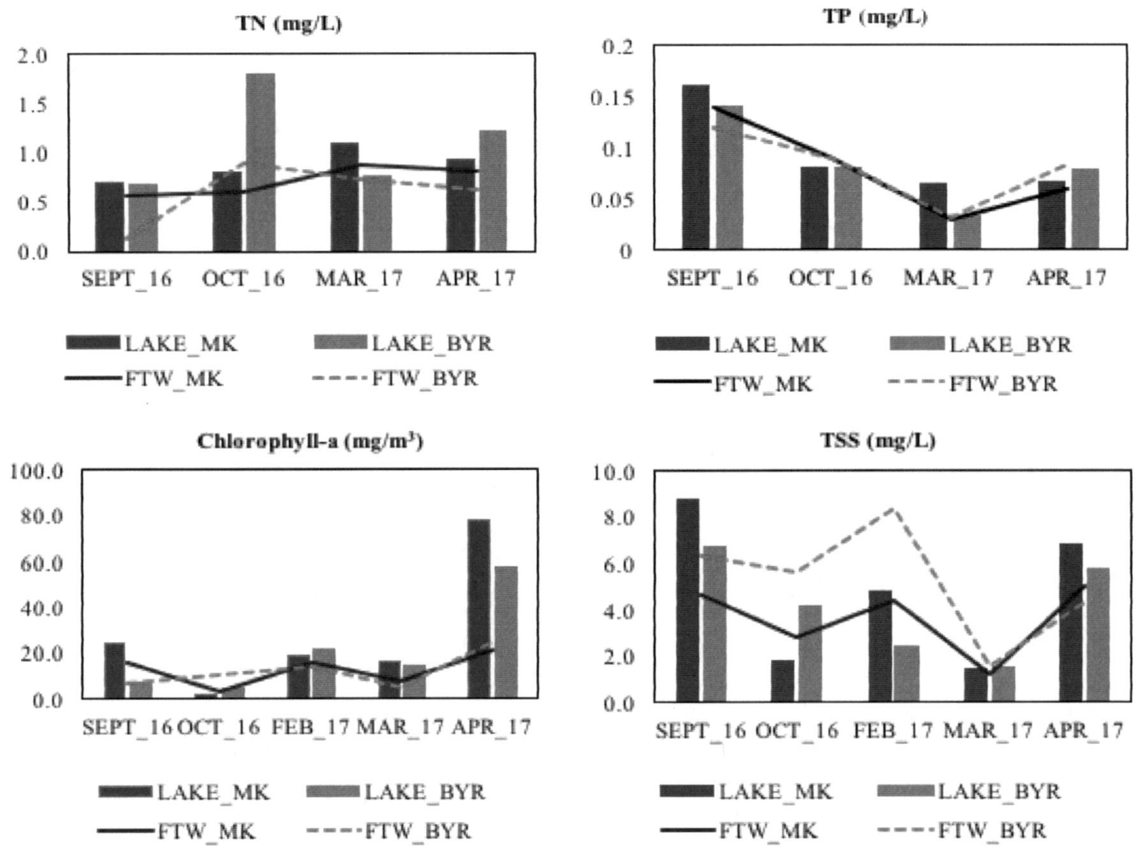

Figure 4. Temporal variation of total nitrogen (TN), total phosphorus (TP), chlorophyll-a and total suspended solids (TSS) concentrations in the lake water and the water inside the FTW system at two locations.

3.2. FTW vegetation stability and Plant growth

Evaluation of the application of FTWs in a big lake should include the sustainability of the FTW frame and the vegetation growth. During anomaly weather in the rainy and transition seasons, heavy rain accompanied by strong wind will cause the turbulence and strong waves in the lake studied. In these conditions the lake will experience mixing upwelling subsequently causing deeper anoxic water column pushed upward to the surface lead to hypoxic lake conditions. During the observation, the lake has experienced these phenomena supported by observed DO level in the lake. Despite experiencing heavy

IOP Conf. Series: Earth and Environmental Science **380** (2019) 012003 doi:10.1088/1755-1315/380/1/012003

waves and lake mixing, the cage and the PVC frame of the FTWs applied remained intact and providing the buoyancy to support the plant growth. The plants in both FTWs grew and adapted well in high wave and even in the hypoxic condition. However, the hypoxic condition did affect the growth of the plants in the FTW_BYR where the area is packed with floating cage aquaculture and has little effect by the water flow. The growth of *C. variegata, C. papyrus* and *E. palaefolius* were slightly slower indicated by the shoot length measured. The plants were less dense in comparison to the plants in the FTW_MK which was influenced by water flow and slightly higher DO level (figures 5,6,7). *C. variegata and C. papyrus* especially grew faster in the FTW_MK and the leaves of *C. variegata* were bigger while *C. papyrus* had higher shoot numbers. *V. zizanioides* also grew well but not *L. flava*. One third of *L. flava* were dead. Results of t test ($p<0.05$) indicated significant different for shoot length and number for *C. variegata*, *C. papyrus* and *E. palaefolius* in the FTW_MK and the FTW_BYR. Nonetheless, the beautiful appearance of plants especially *C. variegata* with bright flowers provided an aesthetic contribution to the lake.

Table 1. Correlation of chlorophyll a, total phosphorus (TP) and total nitrogen (TN).

		CHL	TP	TN
CHL	Pearson Correlation	1	.627**	0.264
	Sig. (2-tailed)		0.009	0.322
	N	16	16	16
TP	Pearson Correlation	.627**	1	0.347
	Sig. (2-tailed)	0.009		0.188
	N	16	16	16
TN	Pearson Correlation	-0.264	-0.347	1
	Sig. (2-tailed)	0.322	0.188	
	N	16	16	16

** Correlation is significant at the 0.01 level (2-tailed).
CHL: chlorophyll-a

IOP Conf. Series: Earth and Environmental Science **380** (2019) 012003 doi:10.1088/1755-1315/380/1/012003

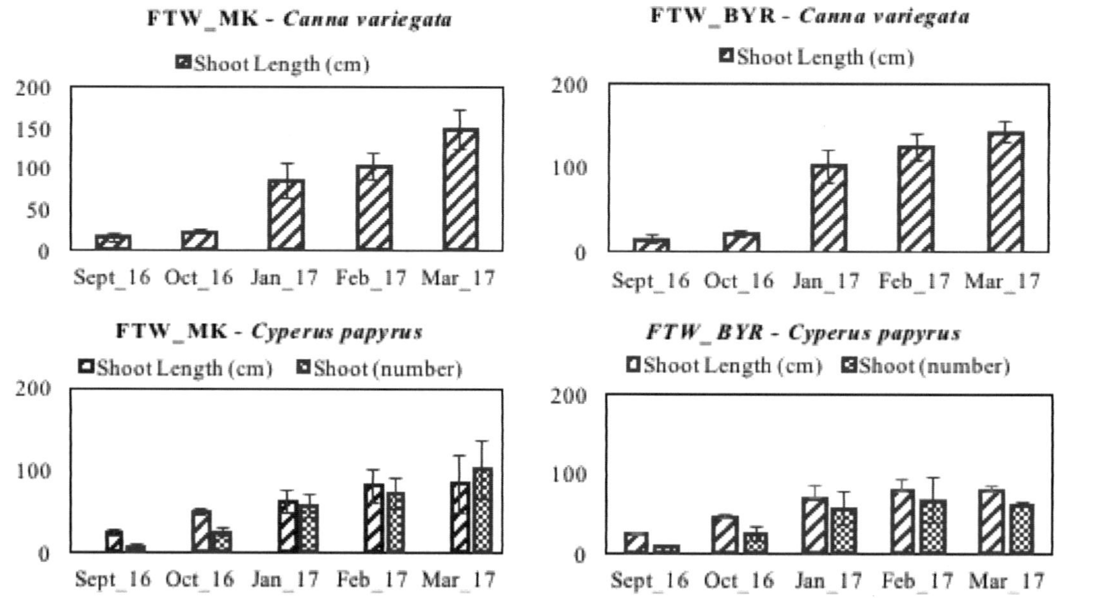

Figure 5. Temporal variation of plant growth for *Canna variegata* (shoot length) and *Cyperus papyrus* (shoot length and number; mean ± standard deviation) at the two FTW systems.

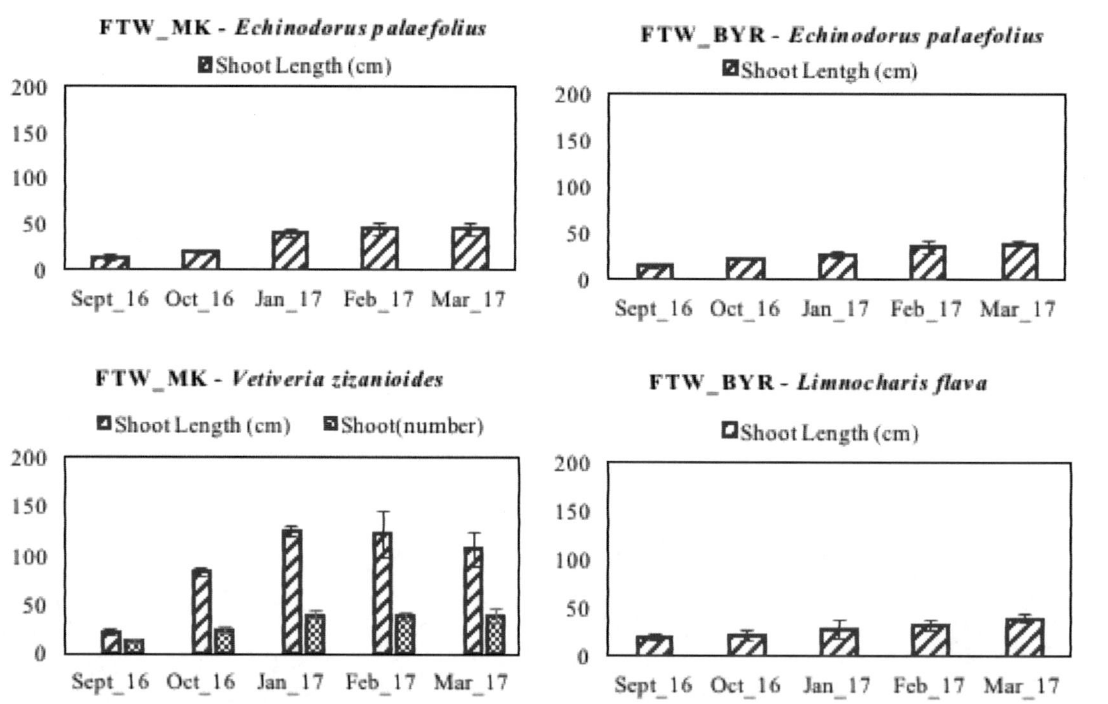

Figure 6. Temporal variation of plant growth for *E. palaefolius* (shoot length), *V.* zizanioides (shoot length and number) and *L. flava* (shoot length) at the two FTW systems.

IOP Conf. Series: Earth and Environmental Science **380** (2019) 012003 doi:10.1088/1755-1315/380/1/012003

Figure 7. Floating Treatment Wetlands at Muko-Muko area (FTW_MK) (left); Comparison of the plant growth for *C. papyrus* (middle) and *C. variegata* (right) at two FTW systems at Muko-Muko (FTW-MK) and Bayur (FTW_BYR) areas after one month period (September – October 2016).

The productivity of plant biomass (shoot part only) were estimated for *C. variegata, C. papyrus* and *V. zizanioides* (table 2). *C. variegata* had the highest biomass productivity. These numbers are much smaller than the numbers estimated by the previous study [17]. They estimated plant productivity to a maximum of 536 g/m^2. d (based on dry weight). This was because in this study a big type of *Cyperus* was used, which had more masses than *C.* papyrus used in this study.

The root length increased from the lowest of 10 – 20 % for *L. flava* to the highest of 40 - 50% for *C. variegata, C. papyrus* and *E. palaefolius* (figure 8), the root masses of *V. zizanioides* and *C. papyrus* were denser than those of *C. variegata* and *E. palaefolius*. The root masses of *L. flava* were the least. Results of t test ($p<0.05$) indicated significant differences for the root length of *C. variegata, C. papyrus* and *E. palaefolius* in the FTW_MK and the FTW_BYR. Our lab scale FTWs study demonstrated better root development and much denser root masses for those plants [22]. The development of the root system of the plants play an important role in the nutrients removal. The root system in the FTWs provide extensive surface area for microbial biofilms attachment and the roots uptake the nutrients while microbial biofilm transforms the nutrients which changes the environmental conditions beneath the mat [14,16].

Table 2. The productivity of FTWs plants.

Plants	Productivity (g/m^2. d)	
Canna variegata	100.28	65.49
Cyperus papyrus	29.72	25.56
Vetiveria zizanioides	20.42	

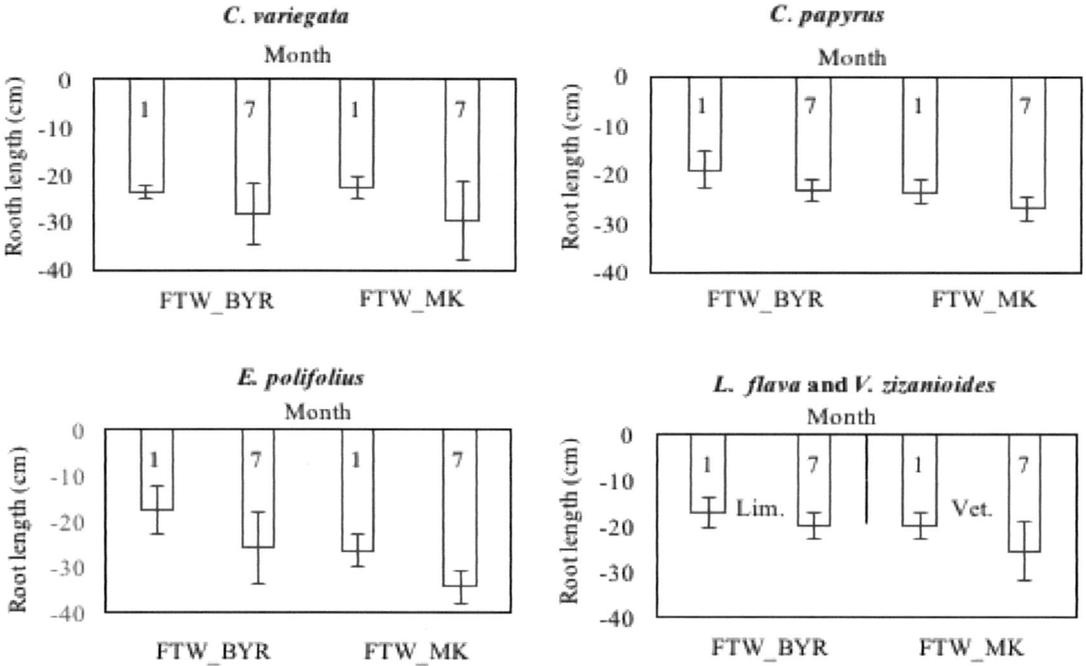

Figure 8. Comparison of the root length for *Canna variegata*, *Cyperus papyrus*, *E. palaefolius*, *Limnocharis flava* (Lim.) and *Vetiveria zizanioides* (Vet.) at the two FTWs systems in September 2016 (Month-1) and in March 2017 (Month-7).

FTWs performance in field studies generally was assessed by determining nutrient uptake by plants. In field studies with large volume and continuous water flow, the lake water constantly mixes and pass by the FTWs much faster than in a controlled batch system giving insufficient time for the pollutant in the water into contact with the network root system of the FTWs. Total N and P accumulated in the plant were calculated from the highest biomass density. The TKN content in the plant ranged from 1.2 – 1.9 g/kg and TP ranged from 0.05 to 0. 18 g/kg. The TKN content in the plants in this study much lower but somewhat similar TP content in the plant with those in the previous study [17]. *C. variegata* had the highest N and P accumulation (table 3).

Total N and P accumulated in plants from this study were quite higher than those reported from other field studies [19]. Assessment of nutrient uptake rate in the FTW field study was usually done by measuring nutrient content in the plant tissues using harvesting strategy of *the* FTW plants in a specified area [17, 19]. Plant harvesting strategy resulted in high nutrient removal by FTWs [17,18,19]. In total, one FTW system in this study could remove > 100 g N and > 7 g P from the lake water. Our study showed that the FTWs in an area that has little effect of flow water (FTW_BYR) had lower N and P uptake and accumulation in plants. During turbulen*t* weather both area where FTWs applied certainly were influenced by a wave and tur*bulent* water in which causing the plants distress and probably affected the nutrient uptake. Tanner and Headley [12] demonstrated that the FTWs in batch trails gave much higher areal removal rate of TN and TP than in flow-through field trials. Unstable lake conditions and often experiencing hypoxic condition influenced the performance of FTWs. Hypoxic condition *can* disrupt pollutant transformation process by aerobic microbial biofilm on the root system [17,19].

Table 3. Nitrogen and phosphorus content in plant shoot.

	N (g/m^2)		P (g/m^2)	
Plants	FTW_MK	FTW_BYR	FTW_MK	FTW_BYR
Canna variegata	35.61	14.95	2.31	2.34
Cyperus papyrus	8.06	6.61	1.17	1.66
Vetiveria zizanioides	8.12		0.38	

3.3. Microhabitat provision based on wildlife observation

The plants, the plant roots and the mats of FTWs show great potential for the creation of microhabitat and the lake restoration. There were few wild plants species observed to be thriving in the new habitat of the FTWs. Beside the grass type of plants there was paddy plant (*Oryza sativa*) spotted in which probably their seeds were dropped by the birds to the FTWs. The plants and the FTW raft appeared to attract some wildlife and the local aquatic life. The fauna spotted at the FTWs were birds, insects, and several types of dragon flies, snails, a snake, frogs, water rats and a big lizard which probably needed a shelter and were looking for food. The flowers of *C. variegata* appeared to attract dragon flies. Small local fish were spotted swimming under the roots, while shrimps and small molluscs were spotted in the area of the roots. At the early monitoring, it was found the fish eggs attached on the palm fibres and also on the roots. During in the hypoxic condition, small local fish population appeared on the surface water inside the FTWs area. At the end of observation, during the plants harvesting, freshwater eels were caught hiding in sugar palm fibres. More than 10 good size of tilapia (15 – 20 cm) were captured inside the net fence which was placed underneath the FTW (section 2.2). Common grass carp were captured in the protective fence of FTWs applied in a eutrophic pond [18].The FTWs clearly could create the habitat which support not only the life of economic local lake's aquatic biota but also other types of terrestrial fauna. Different trophic levels inhabited and frequently visited the FTWs. The FTWs serves as feeding ground or foraging place, a place for reproduction, nursery and a place for shelter. The interaction between trophic levels develop the food web in the FTW ecosystem. Freshwater eels for instance which is the type of animal lives inside the muds, the palm fibres become a suitable media for their habitat and obviously the FTWs provide the food for them as well. Similar studies have been reported on organism inhabited the FTWs [17,18].

Prolonged lake hypoxic condition and frequent algal blooms have decreased the population of economic local fish and molluscs. During these conditions those local fish and molluscs were observed in the FTWs. The presence of FTWs can have benefit to the lake ecosystem by increasing DO level to support aquatic life and creating aerobic conditions for microorganisms responsible in the lake self-purification process [17,19]. In the long term, the FTWs could help maintaining the lake's ecosystem health.

4. Conclusion

Our first application of floating treatment wetlands (FTWs) in a quite big lake gave good evidence that our FTWs system and the plants selected could thrive in the unstable lake conditions such as turbulent water and hypoxic condition. The plants adapted and grew well with good biomass productivity. Improvement of lake water quality in the FTW area may not be apparent but the accumulated nutrients content in the plant shoots suggest high removal of nutrients. In total by the end of observation, one FTWs system could remove > 100 g N and > 7 g P from the lake water. The presence of FTWs will have benefit for the area surrounding not only to maintain better water quality but to provide microhabitat as well for local economic fish, shrimps and molluscs. The FTWs could be developed more in the lake periphery to serve as a conservation area for lake biota. Our study suggests that FTWs can be a promising tool for lake restoration to improve lake water quality and for microhabitat provision. *Canna variegata*, *Vetivveria zizanoides* and *Cyperus papyrus* can be used in FTWs application under a slight stressful environment such as under low DO conditions and turbulent waters.

5. References

[1] Scavia D, Allan D, Arend K K, Bartell S, Beletsky D, Bosch N S, Brandt S B, Briland R D, Daloglu I, DePinto J V, Dolan D M, Evans M A, Farmer T M, Goto D, Han H, Hook T O, Knight R, Ludsin S A, Mason D, Michalak A M, Richards R P, Roberts J, Rucinski D K, Rutherford E, Schwab D J, Sesterhenn T M, Zhang H and Zhou Y 2014 *J. Great Lakes Res.* **40** 226-46
[2] Xu H, Paerl H P, Zhu G, Qin B, Hall N S and Zhu M 2017 *Hydrobiol.* **787** 229-42
[3] Henny C and Nomosatryo S 2016 *IOP Conf. Series: Earth and Environ. Sci.* **31** (2016) 012027
[4] Garg J and Garg H K 2002 *Tropical Ecol.* **43** 355-8
[5] Henny C and Nomosatryo S 2012 *LIMNOTEK* **19** 102-14
[6] Henny C, Santoso AB and Nomosatryo S 2017 *Proc. 16th World Lake Conf. (Bali)* p 557
[7] Syandri H, Junaidi, Mardiah A and Azrita 2017 *Proc. 16th World Lake Conf. (Bali)* p 12
[8] Sulastri, Santoso AB and Nomosatryo S 2017 *Proc. 16th World Lake Conf. (Bali)* p 536
[9] Hackenson L 2005 *Lakes and Reservoir Management.* **10** 71-80
[10] Yeh N, Yeh P, Chang Y H 2015 *Renew. and Sustain. Energy Reviews* **47** 616-22
[11] Zhao F, Xi S, Yang X, Yang W, Li J, Gu B and He Z 2012 *Ecol. Eng.* **40** 53-60
[12] Headley T R and Tanner C C 2006 *Applications of Floating Wetland for Enhanced Stormwater Treatment: A review* Auckland Regional Council, Technical Publication Auckland
[13] Pavlineri N, Skoulikids N Th and Tsihrintzis V A 2017 *Chem. Eng. J.* **308** 1120-32
[14] Tanner C C and Headley T R 2011 *Ecol. Eng.* **37** 474
[15] Jones T G, Willis N, Gough R and Freeman C 2017 *Ecol. Eng.* **99** 316-23
[16] Hartshorn N, Marimon Z, Xuan Z, Cormier J, Chang N B and Wanielista M 2016 *Chemos.* **144** 408-19
[17] Olguin E J, Sanches-Galvan G, Melo F J, Hernandez V J and Gonzalez-Portela R E 2017 *Sci. Tot. Environ.* **584-585** 561-71
[18] Wang C Y, Sample D J and Bell C 2014 *Sci. Tot. Environ.* **499** 384-93
[19] Wang C Y, Sample D J, Day S D and Grizzard T J 2015 *Ecol. Eng.* **78** 15-26
[20] APHA 2005 *Standard methos for the examination of water ans wastewater* Washington DC
[21] Carlson R E and Simpson J 1996 *A coordinator's guide to volunteer lake monitoring methods* North American Lake Management Society 96 pp
[22] Riky K and Henny C 2018 *Proc. Limnol. Soc. Indonesia Annual Meeting (MLI) (Cibinong-Bogor)* **3** p 9

Acknowledgements
This study was supported in part of internal project grant by the Research Centre for Limnology (RCL) which was funded by Indonesian Government. The authors would like to thank B. Teguh Sudiono for technical assistance during FTW construction, Rosidah and the staff at RCL Laboratory for water chemistry for their assistance in the water sample analyses.

Invasive Aquatic Plant Species of Chenderoh Reservoir, Malaysia and Jatiluhur Reservoir, Indonesia

S N Ismail[1*], L Subehi[2], A Mansor[1,3] and M Mashhor [1]

[1]School of Biological Sciences, Universiti Sains Malaysia, 11800 USM, Penang, Malaysia.
[2]Research Centre for Limnology, Lembaga Ilmu Pengetahuan Indonesia, Cibinong, Indonesia
[3]Center for Marine and Coastal Studies, Universiti Sains Malaysia, 11800 USM, Penang, Malaysia

**Corresponding author: norasikin1023@yahoo.com*

Abstract. The sampling was carried out in two sampling sites; Chenderoh Reservoir, Malaysia and Jatiluhur Reservoir, Indonesia. These reservoirs are the initial of cascade reservoirs on Perak River and Citarum River. The aquatic plants distribution and abundance were recorded, and water quality parameters were measured. A total of 23 aquatic plant species from 14 families were identified. The populations of water hyacinth (*Eichhornia crassipes*) and giant salvinia (*Salvinia molesta*) were recorded in all sampling points of both reservoirs. Based on the TSI (Chl *a*), Chenderoh Reservoir was subjected to a mesotrophic, whereas, Jatiluhur Reservoir was subjected to an eutrophic reservoir. *E. crassipes* and *S. molesta* which were abundant in those reservoirs were listed in the 100 world worst invasive alien species.

1. Introduction

At present, 5% of aquatic plants found belong in the 100th of the world's worst invasive alien species list [1], such as water hyacinth (*Eichhornia crassipes*), water fern (*Salvinia molesta*) and a few more to be named. These aquatic plant species could impede the fishing and water transportation activities by forming dense mats on the surface, clogging waterways, blocking out sunlight, and outcompete native floating plants for nutrients. Invasive alien species are regarded as the second largest threat to biodiversity after habitat loss. The special characteristic which makes these invasive alien species successfully invade a given area is the phenotype plasticity which is the ability to alter growth form to suit current condition [2,3].

The expanding global trade and cross boundaries activities allowed the transport of species to various parts of the world including Malaysia and Indonesia. These countries provide areas that are highly favoured by living organism to thrive and flourish in humid and warm tropical region. Some species are colonized and naturalized well to the local environment, while some become invasive. In Malaysia, aquatic plant species considered as noxious weeds are water hyacinth, *Lemna* and hydrilla, [4,5] and the most important species of invasive aquatic plants in Indonesia are water hyacinth, hydrilla, giant salvinia, giant mimosa and water lettuce [6,7]. These alien species easily established in this new environment, and spread so rapidly that native species were sometimes suppressed. This indicates that these plants possess a high degree of adaptation when accompanied with a lack of natural predators. Another characteristic of these plants is their rapid reproduction in both vegetative and generative states.

Annually, invasive species such as water hyacinth costs economic loss of estimated USD $1.4 trillion worldwide [8,9]. The introduction would cause major threats to the native species as they often become subjected to great competition for resources, space, disease and many more. Updated data is required in order to effectively control and manage the spread of invasive species in the ASEAN region. Hence, this study was carried out to determine the distribution and ecology of invasive aquatic plant species in tropical reservoirs, with special reference to Chenderoh and Jatiluhur.

2. Materials and methods

The study was conducted at two sites, Chenderoh Reservoir, Malaysia (figure 1) and Jatiluhur Reservoir, Indonesia (figure 2). The first sampling was carried out in April 2017 at Chenderoh Reservoir and the second sampling was in April 2018 at Jatiluhur Reservoir. Three randomly selected sampling points were deployed in each reservoir for vegetation and water quality survey. The coordinate for each sampling points is shown in Table 1.

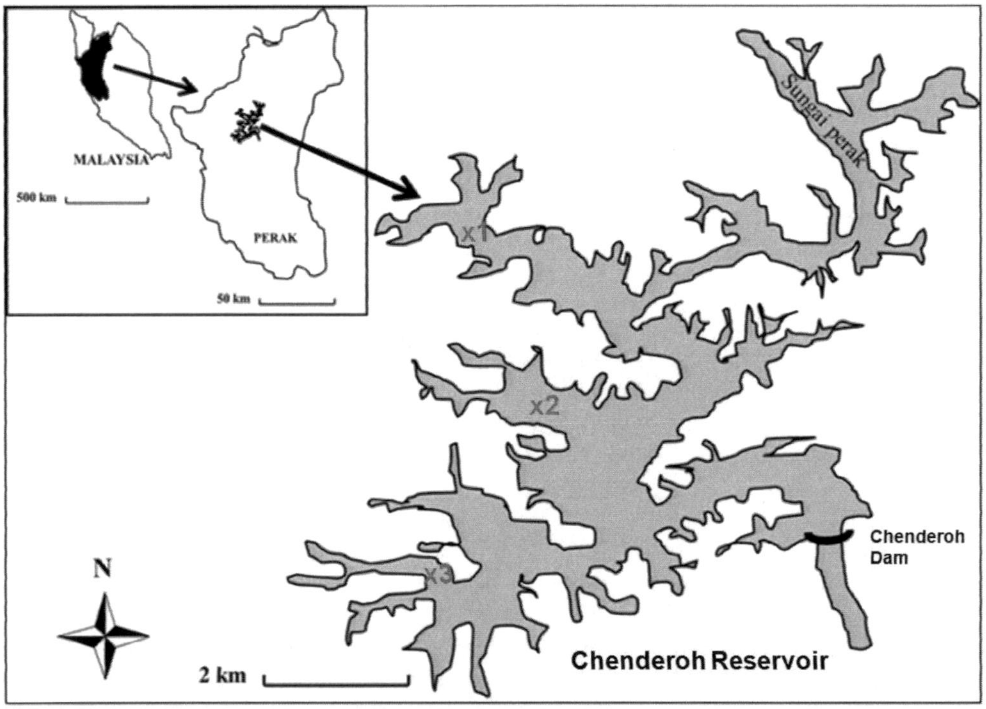

Figure 1. Location of sampling points at Chenderoh Reservoir, Perak, Northern Region of Peninsular Malaysia.

The sampling points are located at the flooded area of the reservoirs except for X1 in Chenderoh Reservoir, which was located at the tributary of Perak River. The depth of each sampling points were recorded with the aid of a depth meter.

The data on aquatic plant distribution and abundance were gathered from a boat according to the surface inventory method [10]. A sampling rake 2.5 meters long was used to collect submersed plant samples. The plants were photographed and identified based on Pancho and Soerjani, 1978 [6]. Plant cover were estimated from the surface based on species presence at sampling points, with species occurrence of 0%-none, 1% to 25%-rare, 26% to 65%-moderate and 66% to 100%-abundant. The invasive species were referred to the IUCN checklist for their ranking in the 100 of the world's worst invasive alien species [11].

The physico-chemical parameters of pH, DO and conductivity were measured in-situ by using portable meters and secchi depth was measured by using a secchi disc. Water samples at the surface were collected with a Van Dorn water sampler and transferred into 500 ml polyethylene bottles. Each sampling point included two replicate samples that were stored in an ice box before transported back to USM Ecology Laboratory, Penang and Research Centre for Limnology Laboratory in LIPI, Indonesia.

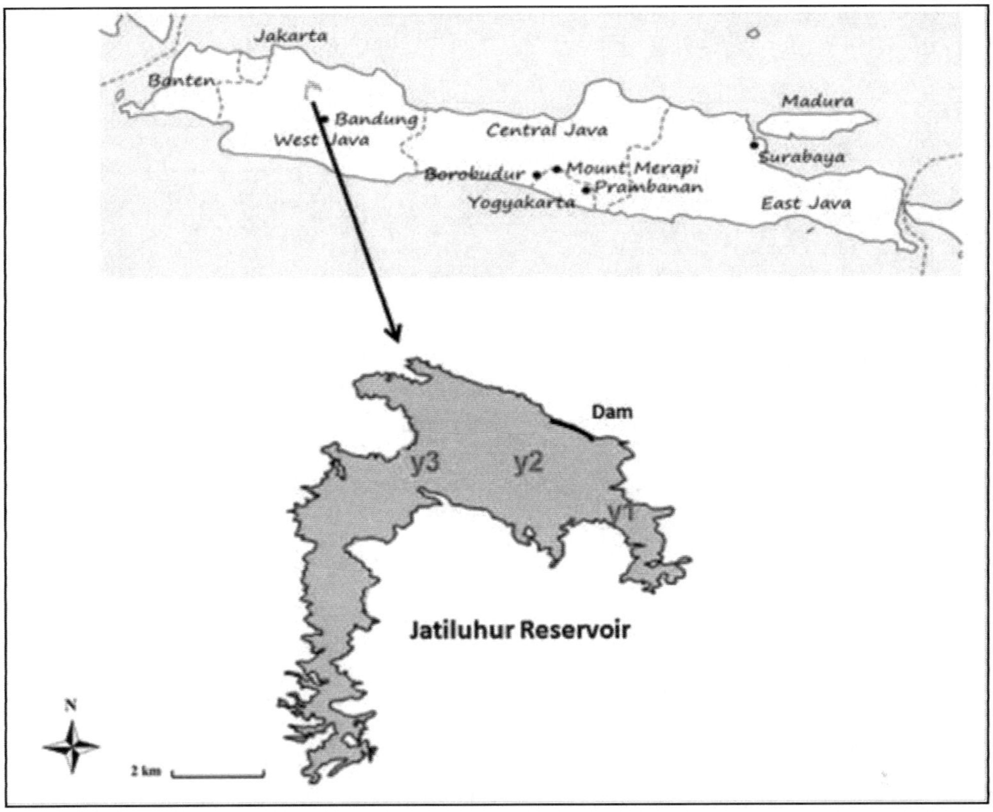

Figure 2. Location of sampling points at Jatiluhur Reservoir, Purwakarta Regency, West Java Province, Indonesia.

Table 1. The coordinates of sampling points in Chenderoh Reservoir and Jatiluhur Reservoir.

Sampling site	Sampling point	Latitude	Longitude	Depth (m)
Chenderoh	X1	4°59'57.78" N	100°56'49.07" E	6
Reservoir	X2	4°58'40.79" N	100°56'56.79" E	3.7
	X3	4°57'50.56" N	100°56'43.14" E	4.8
Jatiluhur	Y1	6°33'02.80" S	107°23'29.81" E	35.4
Reservoir	Y2	6°31'31.10" S	107°21'59.23" E	54.2
	Y3	6°31'54.36" S	107°19'37.22" E	67.4

In the laboratory, analysis for Chlorophyll *a* (Chl *a*; mg/L) and Total Suspended Solids (TSS; mg/L) was done based on Adams, 1990 [12]. Chlorophyll *a* analysis was conducted by filtering 200 ml of water samples through a 0.45 μm Cellulose Nitrate Membrane filter paper. The pigment was extracted using 90% acetone and the optical densities were measured at 750 nm, 664 nm, 647 nm, and 630 nm. The results of Chlorophyll *a* pigments were subjected to Carlson's Trophic State Index (TSI) [13]. The total suspended solids were determined by filtering 250 ml of water samples through a 47 mm glass microfiber filter paper (GF/C). The TSS was analysed by the difference of weight before and after filtering and drying the GF/C filter paper.

3. Results and discussion

A total of 23 aquatic plant species were identified from the two reservoirs (Table 2). From the result, 21 species of aquatic plants from 14 families were recorded in Chenderoh Reservoir, while six species of aquatic plants from three families were recorded in Jatiluhur Reservoir.

Table 2. The list of aquatic plants species recorded in Chenderoh Reservoir, Malaysia and Jatiluhur Reservoir, Indonesia.

Family	Species	Class	Chenderoh Reservoir			Jatiluhur Reservoir		
			X1	X2	X3	Y1	Y2	Y3
Amaranthaceae	*Alternanthera philoxeroides* Mart. Griseb	Emergent	*	*	*	-	-	-
Araceae	*Colocasia esculenta* L. Schott	Emergent	*	*	*	-	-	-
	Lasia spinosa L.	Emergent	*	-	*	-	-	-
	Spirodela polyrrhiza L. Schleid	Free-floating	-	-	*	**	**	**
	Lemna minor L.	Free-floating	*	-	*	**	**	**
	Pistia stratiotes L.	Free-floating	-	-	-	**	**	**
Ceratophyllaceae	*Ceratophyllum demersum* L.	Submerged	***	**	***	-	-	-
Convolvulaceae	*Ipomoea aquatic* Forssk.	Emergent	**	*	*	-	-	-
Hydrocharitaceae	*Hydrilla verticillata* L.f Royle	Submerged	**	**	**	-	-	-
	Najas indica Willd. Cham	Submerged	-	*	**	-	-	-
	Ottelia alismoldes L. Pers	Submerged	-	*	**	-	-	-
Leguminosae	*Neptunia oleracea* Lour.	Emergent	***	*	*	-	-	-
Lentibulariaceae	*Utricularia aurea* Lour.	Submerged	-	*	*	-	-	-
Nelumbonaceae	*Nelumbo nucifera* Gaertn.	Emergent	-	**	***	-	-	-
Onagraceae	*Ludwigia hyssopifolia* G. Don	Emergent	*	*	**	-	-	-
Pandanaceae	*Pandanus helicopus* Kurz.	Emergent	*	***	***	-	-	-
Poaceae	*Brachiaria mutica* Forssk.	Emergent	*	**	**	-	-	-
	Phragmites australis Cav. Trin	Emergent	*	*	*	-	-	-
	Saccharum spontaneum L.	Emergent	*	-	*	-	-	-
Polygonaceae	*Polygonum barbatum* L.	Emergent	*	*	**	-	-	-
Pontederiaceae	*Eichhornia crassipes* Mart. Solms	Free-floating	**	*	*	***	***	***
Salviniaceae	*Salvinia molesta* D.Mitch	Free-floating	*	*	**	***	***	***
	Azolla pinnata R.Br	Free-floating	-	-	-	**	**	**

Note: -: none, *: rare, **: moderate, ***: abundant

The most abundant species recorded were *Eichhornia crassipes*, *Salvinia molesta*, *Ceratophyllum demersum*, *Nelumbo nucifera*, *Neptunia oleracea* and *Pandanus helicopus* (plate 1).

Plate 1. Invasive aquatic plant species of Chenderoh Reservoir and Jatiluhur Reservoir. Note: **A**- *Eichhornia crassipes* **B**- *Salvinia molesta* **C**- *Ceratophyllum demersum* **D**- *Nelumbo nucifera* **E**- *Neptunia oleracea* **F**- *Pandanus helicopus*

Based on recorded species found in both reservoirs, *Eichhornia crassipes* and *Salvinia molesta* were listed in the 100 world worst invasive alien species [11]. *E. crassipes* and *S. molesta* are alien to South East Asian countries. These two species originated from South America and south-eastern

Brazil respectively. The introduction of floating species is known to have shading effects on the underlying ecosystem [14]. In Chenderoh and Jatiluhur, water hyacinth and salvinia can be found in all sampling points. The water hyacinth and salvinia was relatively high in abundance at Jatiluhur Reservoir (plate 2).

Plate 2. Water hyacinth (*Eichhornia crassipes*) population was recorded on the surface of Jatiluhur Reservoir, Indonesia.

The uppermost reservoir on Citarum River is Saguling Reservoir. This reservoir is highly polluted with domestic and industrial effluent from the urban areas of Bandung. The reservoir experiences major water quality problems, including excessive growths of floating plants particularly water hyacinth [15]. Water hyacinth is a very fast growing plant, with populations known to double in as little as 12 days [16]. This species creates shading, and crowding of native aquatic plants which dramatically reduces biological diversity in aquatic ecosystems. It still remains as the world's most problematic water weed despite widespread and various approaches to its control [17]. Soerjani *et al.* [18] rated the floating weed species, *E. crassipes* and *S. molesta* among the most noxious in Southeast Asia. They also singled out *E. crassipes* as the most problematic aquatic weed in Southeast Asia. In fact, works on *E. crassipes* in Southeast Asia including Malaysia are well documented [5,19,20,21,22].

In Chenderoh Reservoir, emergent and submersed aquatic plants dominated the sampling area. *Ceratophyllum demersum*, *Nelumbo nucifera*, *Neptunia oleracea* and *Pandanus helicopus* were the most abundant aquatic plant species in Chenderoh. *Hydrilla verticillata* was found to have been moderately distributed in all sampling points at Chenderoh. Two of the species, hydrilla (*H. verticillata*) and coontail (*C. demersum*) are considered as noxious weeds. Although they are considered as weeds, these two species however are being utilized as aquarium plants [4].

The concentration of dissolved solids will affect water conditions especially on the penetration of sunlight into the water. The sunlight penetration into water bodies will be shorter with high content of dissolved solids. This will affect the productivity of phytoplankton that are very dependent on sunlight. Table 3 shows the water quality for Chenderoh Reservoir (x) and Jatiluhur Reservoir (y).

Table 3. Results of water quality analysis at Chenderoh Reservoir, Malaysia and Jatiluhur Reservoir, Indonesia.

Sampling points	SD (m)	DO [mg/l]	pH	Cond. [mS/cm]	TSS (mg/L)	TSI(Chl *a*)
X1	117.5	6.45	6.73	0.022	0.0097	48.0
X2	37.5	7.93	6.33	0.472	0.0132	41.7
X3	135.5	7.66	6.32	0.414	0.0036	38.6
y1	116	8.07	5.8	0.24	0.0016	68.6
y2	115	9.87	6.1	0.23	0.0014	65.3
y3	112.5	8.15	5.7	0.23	0.0010	62.4

The degradation of environment, including in water quality of reservoir occurred in the sampling areas. Anthropogenic activities contributed to the decrease of water quality. The growth of population influenced the activities of people in the surroundings of the reservoirs. Based on the results of Carlson's TSI (Chl *a*) analysis, Chenderoh Reservoir can be considered as mesotrophic, whereas, Jatiluhur Reservoir can be considered as an eutrophic reservoir. The TSI (Chl *a*) in Chenderoh is in mesotrophic ranges (TSI 40-60) [23]. The trend of macrophyte problems were seen at Jatiluhur Reservoir due to excessive nutrient load. The excessive amount of toxic chemicals was being dumped into the Citarum River by industries and housing area. Not surprisingly, since 2008, nearly 60 percent of the river's fish species have been destroyed. Citarum River is listed as the world's most polluted river [24].

4. Conclusion

A few invasive aquatic plants species were identified at Chenderoh and Jatiluhur reservoirs. The species were free-floating, submersed and emergent plants. The differences of aquatic plant class distribution in these two reservoirs are related to the reservoir morphology and depth. Submerged and emergent species could be found in a shallow, flow through reservoir and free-floating species were found on the surface of lakes and reservoirs. The invasive species are thriving well due to favourable condition of water quality. The special phenotype plasticity characteristic supports the plants to be able to alter growth form to suit current condition.

5. References

[1] Luque G M, Bellard C, Bertelsmeier C, Bonnaud E, Genovesi P, Simberloff D, and Courchamp F 2014. The 100th of the world's worst invasive alien species. *Biological invasions*, *16*(5), 981-985.

[2] Dewitt T J and Scheiner S M 1994. *Phenotypic plasticity: Functional and conceptual approaches*. Oxford, UK: Oxford University Press.

[3] Nicolaus M and Edelaar P 2018. Comparing the consequences of natural selection, adaptive phenotypic plasticity, and matching habitat choice for phenotype–environment matching, population genetic structure, and reproductive isolation in meta-populations. *Ecology and evolution*, *8*(8), 3815-3827.

[4] Mashhor M 1987. The Major Aquatic Plants of Peninsular Malaysia. *Aquatics*, 9(1), 17-19.

[5] Mashhor M 1996. Noxious floating weeds of Malaysia. *Hydrobiologia*, *340*(1-3), 121-125.

[6] Pancho J V and Soerjani M 1978. *Aquatic weeds of Southeast Asia: a systematic account of common Southeast Asian aquatic weeds.* National Publishing Cooperative Incorporated. Quezon City, Philippines. 130 pages.

[7] Tjitrosoedirdjo S S 2005. Inventory of the invasive alien plant species in Indonesia. *BIOTROPIA-The Southeast Asian Journal of Tropical Biology*, (25).

[8] Pimentel D, Zuniga R and Morrison D 2005. Update on the environmental and economic costs associated with alien-invasive species in the United States. *Ecological economics*, *52*(3), 273-288.

[9] Bradshaw C J, Leroy B, Bellard C, Roiz D, Albert C, Fournier A and Courchamp F 2016. Massive yet grossly underestimated global costs of invasive insects. *Nature communications*, *7*, 12986.

[10] Parsons J K 2001. *Aquatic plant sampling protocols*. Washington State Department of Ecology, Environmental Assessment Program.

[11] Global Invasive Species Databade 2018. Downloaded from http://www.iucngisd.org/gisd/100_worst.php on 19-09-2018

[12] Adams D V 1990. *Water and Wastewater Examination Manual*. Lewis Publisher, Chelsea. 1st edition. 264 pages

[13] Carlson R E 1977. A trophic state index for lakes. *Limnology and Oceanography*. 22(2): 361-369.

[14] Caraco N F and Cole J J 2002. Contrasting impacts of a native and alien macrophyte on dissolved oxygen in a large river. *Ecological Applications*, **12**, 1496– 1509.

[15] Hart B T, van Dok W, and Djuangsih N 2002. Nutrient budget for Saguling reservoir, West Java, Indonesia. *Water Research*, *36*(8), 2152-2160.

[16] Lowe S, Browne M, Boudjelas S, and De Poorter M 2000. *100 of the world's worst invasive alien species: a selection from the global invasive species database* (Vol. 12). Auckland: Invasive Species Specialist Group.

[17] Heard T A and Winterton S L 2000. Interactions between nutrient status and weevil herbivory in the biological control of water hyacinth. *Journal of Applied Ecology*, *37*(1), 117-127.

[18] Soerjani M, J V Pancho and N V Voung 1975. Aquatic weed problems and control in Southeast Asia. *Hyacinth Control Journal*. 13: 2-3.

[19] Little E C S 1969. *The Floating Island of Rawa Pening*. PAN, 15 pp.

[20] Nguyen V V 1973. Report on the aquatic weed problem of the Brantas River multipurpose project. Biotrop /WR/73/654. 10 pp.

[21] Soerjani M 1979. Recent trend in aquatic weed management in Indonesia. Proceeding Asian-Pacific Weed Science Society; Seventh Conference. Supp. Vol. 12 pp.

[22] Lim W C and A Salleh1983. *Eichhornia crassipes:* a serious weed in the Muda Irrigation Area. Proceeding of Weed Science in the Tropic Rajan A. & M. Rosli (eds): 12-18.

[23] Sharip Z, Zaki A T, Shapai M A, Suratman S and Shaaban A J 2014. Lakes of Malaysia: Water quality, eutrophication and management. *Lakes & Reservoirs: Research & Management*, *19*(2), 130-141.

[24] Dikayana T and Muhammad Z R 2018. *Indonesia's Citarum: The World's Most Polluted River*. The Diplomat, April 28, 2018 https://thediplomat.com/2018/04/indonesias-citarum-the-worlds-most-polluted-river/

The River Basin Spatial Informative Nesting (E-BASIN) framework as an alternative approach for flood disaster management

M Y Izham[1*], R Aznarahayu[1], M A Nurul Azni[1] and M N Norashila[1]

[1]School of Distance Education, Universiti Sains Malaysia, Malaysia, 11800 USM, Penang, Malaysia.

**Corresponding author: izham@usm.my*

Abstract. The usage of technology is vital towards flood management. It helps on delivering the information on disasters occurred in the affected areas. Heavy rainfall and flood disaster occur frequently in Malaysia during the monsoon seasons. This causes significant impact towards human life, loss of properties and damage to the surrounding environment. The requirement on effective flood disaster management within river basin boundaries has become a necessity by managing the weather records, simulating present and future flood events; and providing best approach of reducing flood impact assessment before, during and after. This article aims to propose a new framework in flood disaster management as an alternative way to disseminate flood events, namely E-Basin framework. E-Basin is designed via "light and easy" approach to visualise, alert and report flood events among affected local residents, societies and related agencies situated in flood-prone areas. The implementation of E-Basin framework could provide further guideline among the involved agencies for improving effectiveness of flood disaster management geographically in the 4th Industrial Revolution era.

1. Introduction

Flood is considered a major natural hydrological disaster, causing significant impact towards human life, loss of properties and surrounding environment [1-4]. Frequent flood events would contribute negative impacts towards development planning, transportation access, buildings and other properties [5]. Since the 1920s, floods have been reported as a common occurrence in many countries. Previous studies stated that 9% of the total land area and 22% of the total population were directly affected by floods [6-10]. According to [11], over the recent decade, the frequency of floods disaster occurrence has increased. Moreover, the countries are located in the Asia and Pacific are the most badly affected states by the flood including Malaysia [12].

Malaysia is classified as having an equatorial climate with high humidity and experiencing abundant and continuous heavy rainfall every year [13] with the average of approximately 2,500 mm a year in Peninsular Malaysia, 3,000 mm a year in Sabah and 3,500 mm a year in Sarawak [7,10,14]. The amount of rainfall shows that Malaysia has potential to flood disaster exposure and water overflows every year [8]. In Malaysia, there are two basic types of floods that have been classified, namely flash flood and monsoon flood [15]. Flash flood is a fluvial type of flooding triggered by causative rainfall, and the drainage system not able to flow well as a result of the heavy rainfall in a short time [16]. Flash flood may possibly occur very quickly because the water can only move via streams or narrow channels and the drainages are unable to accommodate for any sudden high-water flow. This has been observed in busy cities as a result of human activities such as infrastructure development near the river areas as well as uncontrolled littering causing clogged drains and waterways [8,17]. However, it takes a short period to recede to the normal water level compared to the monsoon flood. For instance, flash flood disaster which occurred in Cameron Highlands on 23rd October 2013 claimed 3 lives, injuries, heavy mudflows and damage of properties [18]. Based on the interviews carried out by researchers among the victims, they concluded that the early warning system was not operating properly, and safety response requirements and quick evacuation access failed.

Moreover, monsoon flood occurs during two wind seasons that contributes to heavy rainfall, which is from May to August (The Southwest Monsoon) and from November to February (The Northeast Monsoon) [19,10]. Significant rainfall intensity within several hours would generate flash floods, causing damages to infrastructure, contamination of water supplies and harvest losses [20]. Many countries such as Indonesia, west Malaysia, Southern Thailand and Sri Lanka faces significant flood hazards in the year of 2014 due to the northeast monsoon [21]. Malaysia is among the countries with the worst affected area in several states, with more than 200,000 people affected, and 21 people killed due to the monsoon [22]. Researchers predict that frequent catastrophic floods will occur in the future [23]. Thus, related government agencies and communities have begun to enhance current flood models, inventing green technologies, attempting cost effective strategies to reduce negative impacts on living things, properties, infrastructures, agriculture, human health and economic status [5,19,23,24,25,26].

Several organizations initiated positive programs and invented prediction models to deal with the problem. For instance, there are five strategies executed by the Disaster Management and Relief Committee of Malaysia on the development of flood management programs such as prevention, protection, preparedness, emergency actions and recovery [27]. Nevertheless, the policy on information sharing and data compatibility is still disseminated among researchers and government agencies, without involving local communities. In fact, the needs of local community involvements are crucial towards verifying actual flooded areas spatially, identifying the extent of damage in the affected areas within river basin and to provide evacuation guidelines effectively. Moreover, current flood management systems do not integrate the involvement of local societies to participate, report and simulate flood event due to the designation of different users. Hence, the more protections and new legislation need to be reviewed in accordance with such requirements for urban and rural areas.

In the flood's scenario, several agencies have begun getting attention on enhancing the use of geospatial information [28]. Geospatial information is a very useful tool for geospatial analysis during flood mitigation processes. There are several usages of geospatial information including position, geospatial distribution, location clustering and status of damages related with hazard [29]. Geospatial information must always be in a good condition and must speed up the action towards predicting the possible flooded areas [30]. However, the local community and a few agencies still lack awareness regarding river basin. River basin is important for giving a successful flood prediction, preparedness and post-flood execution. Hence, a holistic approach to flood management needs to be executed by engaging the local community, related government agencies and researchers within a river basin scope [8,31].

Flood disaster that occur at river basin boundary have been an unavoidable problem in many cities around the world. The damages of properties from the flood occurrence provide limitations to the varieties of sectors, either human activities, health or economic issues. In relation to water-resource planning and management mainly in disaster mitigation, decision makers often rely on flood studies, which usually use numerical models to simulate flood inundation. Early warning systems and flood risk mapping are examples of the advantages of flood models [32,33].

Research has been done in terms of developing flood models, environmental assessment and early warning systems; but the issues on information sharing policies and data compatibility cause uncertainty in terms of predicting flood areas, damage, evacuations and executions within a limited time. In addition, current flood management models focus on managing flood disaster from the administration's perspective and only involve the agencies responsible for managing flood disaster such as government agencies and NGOs. Those models are limited in terms of public or community involvement. However, according to [34], community or public engagement is crucial for a more effective and efficient flood disaster management.

2. E-Basin for Managing Flood Disasters

In this study, the E-Basin framework is designed as an alternative tool for delivering spatial information via web application. It is as a medium of delivering information for saving lives and reducing damages due to the flood event. The aims of E-Basin are to engage the community, the involved authorities and experts to enhance their responsibilities and collaboration related to managing of flood disasters. Through E-Basin, users can share the information, access the right information and obtain updates easily on news about flood disasters. Generally, this application will design for self-study within smartphone application and personal computers and users can access it anytime and anywhere.

E-Basin differs from the other flood prediction system by easing data transmission within a short period of time. Therefore, it enables users to access the information on the possible flood areas in quick time. In E-Basin, the users would be able to retrieve and update information on current weather events. The community will retrieve the news quickly and take precautions before flood occurs. As stated by [35], the management of flood disasters need to access a vital infrastructure system that is able to provide the right information and timely warnings to assist the community respond about flood events. As a result, this initiative has been implemented to provide great insights into the development of flood.

The main focus of this research work is the designation aspect of E-Basin framework and initial spatial information inventory of flood conditions and river communities. E-Basin encourages community involvement as it allows the user to send the information of flood to the system. It enhances interaction of various involved agencies and system with the affected population and the situation of limited transmission data can be overcome. This will lead to an easy and quick distribution on flood-related information sharing process. With the implementation of E-Basin, the system enables the community to obtain quick references for flood preparedness, evacuation route, access to the nearest help centre, domestic equipment and medical attention.

In accordance with statement of [27], the involved agencies in flood management, i.e. the government will only act after the disaster occurred and overlooked the policy aspect as the preparation for the future. E-Basin would act as a medium in educating "My river basin - my homeland" concept towards the bottom billion by providing the facilities to the society so that they more alert and aware on sustaining natural environmental existence. Most importantly on E-Basin framework is expected to reduce the redundancy of spatial informative delivery and usage among government agencies, which provides a holistic, coordinated, enriched and complete guide towards centralized geospatial natural disaster information centres. Additionally, E-Basin leads to sustainable economic planning, activities and recovery and introduces a more superior, organized system while minimizing conflicts that may exist during the disaster period.

3. Materials and methods

The purpose of this research is to design an E-Basin framework as an additional tool to inform the flood disasters in any area. In this study, the researchers use a river basin extended as the main centre of distributing geographical information. The main focus of this study is on the aspect of E-Basin framework, designing and initial spatial information inventory of flood conditions and river communities. The E-Basin system is embedded within desktop and smart phone so that the users could promptly access the information on floods quickly. The study area is in the Sungai Perak river basin, located in Perak, Malaysia. Sungai Perak is located in the Northern Peninsular Malaysia, which obtain frequent heavy rainfalls and records high water level in the river. Thus, resulting in high vulnerability to the entire Sungai Perak river basin due to frequent floods.

The overall methodology of the study is divided into three phases; the preliminary study (first phase), constructing the E-Basin framework (second phase) and testing the E-Basin framework (third phase). Figure 1 depicts the research activities related in the study.

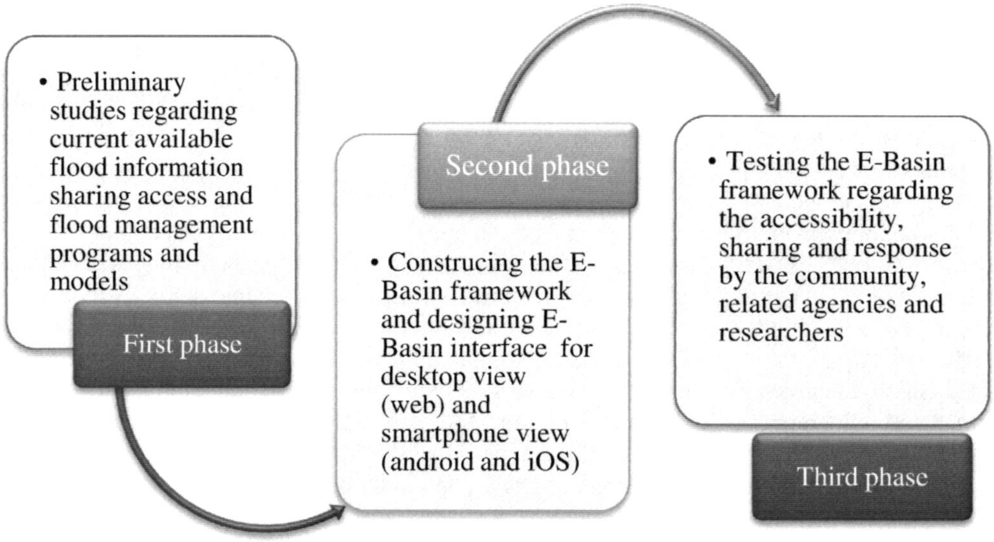

Figure 1. Research Activities.

3.1. First phase

The first phase involves analysing the data on current flood information before disseminating to the users. The data is available to be accessed and shared among researchers, the involved agencies and communities. In E-Basin, the end users (i.e. local community) would be able to simulate predicted flood events and taking part towards observing the river basin.

3.2. Second phase

The second phase involves installation and setup of the server, client server and constructing E-Basin framework. The spatial layers identified in phase one will be captured using ArcGIS, processed and generating flood plain areas via coupling with Hydrologic Engineering Centre-Hydrologic Modelling System (HEC-HMS) simulation model. This phase will then continue with the designing of E-Basin web interface for viewing in desktop and smartphone (android and iOS) application. The system is embedded with spatial results sharing, prompt information updating, flexible expansibility and user response.

3.3. Third phase

In the third phase, the spatial information access via desktop and smartphone will be tested in terms of E-Basin continuous update of flood spatial information and accessibility among the local communities, related agencies and researchers as well as E-Basin information retrieval from these users. The E-Basin framework will be validated with the current available flood information systems.

4. Results and discussion

E-Basin framework is the first research to integrate smart of spatial informative delivery for flood prediction incorporating fundamental aspect of geospatial information for fully geospatial based integration. E-Basin involves related agencies on managing flood issues, researchers and local communities. The E-Basin is implemented as a web-based application to allow users in updating and viewing information regarding flood events. This application is expected to provide the best foundation of participation from the communities via prompt update of possible accurate flood spots and current weather events. Therefore, the communities will be more prepared on the upcoming flood disasters, thus reducing the loss of life and economic losses. E-Basin attempts to centralize the system of flood management, which involvement of various communities, agencies and researchers. Through E-Basin system, it would educate the communities to be more alert the surroundings of the river basin and sustaining natural environmental existence via the concept of "My river basin – my homeland". This application will lead to eliminate the redundancy of spatial informative delivery and usage between government and related agencies.

5. References

[1] Curebal, I., Efe, R., Ozdemir, H., Soykan, A., & Sönmez, S. (2016). GIS-based approach for flood analysis: Case study of Keçidere flash flood event (Turkey). *Geocarto International*, 31(4), 355–366. doi:10.1080/10106049.2015.1047411.

[2] Kia, M. B., Pirasteh, S., Pradhan, B., Mahmud, A. R., Sulaiman, W. N. A., & Moradi, A. (2012). An artificial neural network model for flood simulation using GIS: Johor River Basin, Malaysia. Environmental Earth Sciences, 67(1), 251–264. doi:10.1007/s12665-011-1504-z

[3] Liu, C., & Li, Y. (2016). GIS-based dynamic modelling and analysis of flash floods considering land-use planning. *International Journal of Geographical Information Science*, 31(**3**), 1–18. doi:10.1080/13658816.2016.1207774

[4] Siahkamari, S., Haghizadeh, A., Zeinivand, H., Tahmasebipour, N., & Rahmati, O. (2017). Spatial prediction of flood-susceptible areas using frequency ratio and maximum entropy models. *Geocarto International*, 1–15. doi:10.1080/10106049.2017.1316780

[5] Elias, Z., Hamin, Z., & Othman, M. B. (2013). Sustainable management of flood risks in Malaysia: Some lessons from the legislation in England and Wales. *Procedia - Social and Behavioral Sciences*, **105**, 491–497. doi:10.1016/j.sbspro.2013.11.052

[6] Aliagha, U. G., Jin, T. E., Choong, W. W., Jaafar, M. N., & Ali, H. M. (2014). Factors affecting flood insurance purchase in residential properties in Johor, Malaysia. *Natural Hazards and Earth System Sciences*, **14**, 3297–3310. doi:10.5194/nhess-14-3297-2014

[7] Khalid, M. S., & Shafiai, S. (2015). Flood disaster management in Malaysia: An evaluation of the effectiveness flood delivery system. *International Journal of Social Science and Humanity*, 5(4), 398–402. doi:10.7763/IJSSH.2015.V5.488

[8] Othman, M., Ahmad, M. N., Suliman, A., Arshad, N. H., & Maidin, S. S. (2014). COBIT principles to govern flood management. *International Journal of Disaster Risk Reduction*, **9**, 212–223. doi:10.1016/j.ijdrr.2014.05.012

[9] Salleh, N. 'Aaina, Mustaffa, C. S., & Ariffin, M. T. (2013). Proposing instrument to measure impression management among flood victims. *International Journal of Social Science and Humanity*, 3(**6**), 538–542. doi:10.7763/IJSSH.2013.V3.299

[10] Tan, M. L., Ibrahim, A. L., Duan, Z., Cracknell, A. P., & Chaplot, V. (2015). Evaluation of six high-resolution satellite and ground-based precipitation products over Malaysia. *Remote Sensing*, 7(**2**), 1504–1528. doi:10.3390/rs70201504

[11] Kourgialas, N. N., & Karatzas, G. P. (2011). Flood management and a GIS modelling method to assess flood-hazard areas—a case study. *Hydrological Sciences Journal*, 56(**2**), 212–225. doi:10.1080/02626667.2011.555836

[12] Hiwasaki, L., Luna, E., Syamsidik, & Shaw, R. (2014). Process for integrating local and indigenous knowledge with science for hydro-meteorological disaster risk reduction and climate change adaptation in coastal and small island communities. *International Journal of Disaster Risk Reduction*, **10**, 15–27. doi:10.1016/j.ijdrr.2014.07.007

[13] Kuok, K. K., Liew, Z. Z., & Chiu, P. (2013). Flood map development by coupling satellite maps and three-dimensional drafting software: Case study of the Sarawak River Basin. Water SA, 39(**1**), 175–182. doi:10.4314/wsa.v39i1.18

[14] Tahir, P. A. (2009). Flood management in Malaysia. In Flood Hazard Mapping Seminar (pp. 1–28). Manila, Philipines. Retrieved from http://www.icharm.pwri.go.jp/training/2009seminar/3_paridah_malaysia.pdf

[15] Garba, S., & Iya, D. (2015). FLOODS IN MALAYSIA. Historical Reviews, Causes, Effects and Mitigations Approach, (November 2014).

[16] Nizam, M. S & Naser, A. (2017). An overview of road damages due to flooding: Case study in Kedah state, Malaysia. Proceedings of the International Conference of Global Network for Innovative Technology and AWAM International Conference in Civil Engineering (IGNITE-AICCE'17), (October 2017).

[17] Sipon, S., Sakdan, M. F., Mustaffa, C. S., Marzuki, N. A., Khalid, M. S., Ariffin, M. T., … Abdullah, S. (2015). Spirituality among flood victims: A comparison between two states. Procedia - Social and Behavioral Sciences, 185, 357–360. doi:10.1016/j.sbspro.2015.03.410

[18] Khalid, M. S., Mustaffa, C. S., Marzuki, M. N., Sakhdan, Sipon, S., Ariffin, M. T & Shafiai, S. (2015). Failure to React Positively to Flood Early Warning Systems: Lessons Learned by Flood Victims from Flash Flood Disasters: The Malaysia Experience. *International Journal of Social, Behavioral, Educational, Economic and Management Engineering*, 9(**5**), 1316-1320.

[19] Austin, O. C., & Baharuddin, A. H. (2012). Risk in Malaysian agriculture: The need for a strategic approach and a policy refocus. Kajian Malaysia, 30(**1**), 21–50.

[20] Teong, K. V., Sukarno, K., Hian, J., & Chang, W. (2017). The Monsoon Effect on Rainfall and Solar Radiation in Kota Kinabalu, 4(4), 460–465.

[21] Akasah, Z. A., & Doraisamy, S. V. (2015). 2014 Malaysia flood : impacts & factors contributing towards the restoration of damages, 2(**14**), 53–59.

[22] Basri, A. B., Ismail, A. F., Khairolanuar, M. H., Sobli, H. M., Badron, K., & Hasan, M. K. (2016). Analyses of Rainfall Rate during Malaysian 2014 Flood Event. 11(8), 237–246.

[23] Shah, M. A. R., Rahman, A., & Chowdhury, S. H. (2015). Challenges for achieving sustainable flood risk management. Journal of Flood Risk Management, 1–7. doi:10.1111/jfr3.12211

[24] Abram, N. K., Xofis, P., Tzanopoulos, J., MacMillan, D. C., Ancrenaz, M., Chung, R., Knight, A. T. (2014). Synergies for improving oil palm production and forest conservation in floodplain landscapes. PLoS ONE, 9(6), 1–12. doi:10.1371/journal.pone.0095388

[25] Hammond, M. J., Chen, A. S., Djordjević, S., Butler, D., & Mark, O. (2015). Urban flood impact assessment: A state-of-the-art review. *Urban Water Journal*, 12(**1**), 14–29. doi:10.1080/1573062X.2013.857421

[26] Klijn, F., Mens, M. J. P., & Asselman, N. E. M. (2015). Flood risk management for an uncertain future: economic efficiency and system robustness perspectives compared for the Meuse River (Netherlands). Mitigation and Adaptation Strategies for Global Change, 20(**6**), 1011–1026. doi:10.1007/s11027-015-9643-2

[27] Shafiai, S., & Khalid, M. S. (2016). Examining of Issues on Flood Disaster Management in Malaysia, 6(**20**), 51–56.

[28] Kamin, M., Ahmad, N. F. A., Razali, S. N. M., Hilaham, M. M., Rahman, M. A., Ngadiman, N., & Sahat, S. (2017). Flood prediction using geographical information system (GIS) application at Sungai Sembrong. *Journal of Telecommunication, Electronic and Computer Engineering*, 9(**2–12**).

[29] Xie, Jibo, Tengfei Yang & Guoqing, L. (2018). Extracting Geospatial Information from Social Media Data for Hazard Mitigation, Typhoon Hato as Case Study. International Conference on Geograpic Information Science, 65(**9**), 1–6.

[30] Fukuzono, T., Sato, T., Takeuchi, Y., Takao, K., Shimokawa, S., Suzuki, I., Ikeda, S. (2006). Participatory Flood Risk Communication Support System (Pafrics). In T. Ikeda, S. Fukuzono, T. and Sato (Ed.), A better integrated management of disaster risks: Toward resilient society to emerging disaster risks in mega-cities. (pp. 199–211). TERRAPUB and NIED.

[31] Amirebrahimi, S., Rajabifard, A., Mendis, P., & Ngo, T. (2016). A framework for a microscale flood damage assessment and visualization for a building using BIM–GIS integration. International Journal of Digital Earth, 9(**4**), 363–386. doi:10.1080/17538947.2015.1034201

[32] Farid, M., Mano, A., & Udo, K. (2012). Urban flood inundation model for high density building area. Journal of Disaster Research, 7(**5**), 554–559.
[33] Syed Hussain, T. P. R., & Ismail, H. (2013). Flood frequency analysis of Kelantan River Basin, Malaysia. World Applied Sciences Journal, 28(**12**), 1989–1995. doi:10.5829/idosi.wasj.2013.28.12.1559
[34] Amendola, A., Linnerooth-Bayer, J., Okada, N., & Shi, P. (2008). Towards integrated disaster risk management: Case studies and trends from Asia. Natural Hazards, 44(2), 163–168. doi:10.1007/s11069-007-9152-z
[35] Sunkpho, J., & Ootamakorn, C. (2011). Real-time flood monitoring and warning system, 33(November 2002), 227–235.

Acknowledgments
The authors would like to express their appreciation to School of Distance Education, Universiti Sains Malaysia for providing the financial support via RUI research grant No.1001/PJJAUH/816293 to carry out this research work.

Water quality simulation for rehabilitation of a eutrophic lake in Selangor, Malaysia

H L Koh[1*], W K Tan[2], S Y Teh[3] and C J Tay[3]

[1] Jeffrey Sachs Centre on Sustainable Development, Sunway University, Bandar Sunway, 47500 Subang Jaya, Selangor, Malaysia
[2] School of Mathematical Sciences, Sunway University, Bandar Sunway, 47500 Subang Jaya, Selangor, Malaysia
[3] School of Mathematical Sciences, Universiti Sains Malaysia, 11800 USM, Penang, Malaysia

**Corresponding author: hocklyek@sunway.edu.my*

Abstract. More than 50 % of lakes worldwide, including 60 % in Malaysia, are contaminated by eutrophication due to excessive nutrient inputs. These nutrients include phosphorus and nitrogen derived from agriculture, industry and domestic wastes. Eutrophication is an undesirable lake condition characterized by turbid water and dominated by abundant growth of algae, resulting in suppressed macrophytes and in reduced ecosystem services. Healthy lakes provide ecosystem services that have been valued to be worth billions to the Malaysian Gross Domestic Production annually. Rehabilitation of eutrophic lakes located in urban areas is therefore urgently needed. A recreational lake located in Selangor Malaysia has shown persistent signs of eutrophication with chlorophyll-a level fluctuating between 10 and 25 μg/L. Left untreated, the lake water quality will deteriorate further, potentially inducing a regime shift to hypereutrophic condition, caused by intense interaction between water column phosphorus and sediment phosphorus. Based upon water and sediment quality data compiled over the last eight years, this paper presents analysis and model simulations for implementation of a sustainable program for controlling eutrophication using low cost technology.

1. Introduction

Eutrophication is a common feature in shallow tropical lakes surrounded by large human settlements. Eutrophic lakes are characterized by turbid water with an abundance of floating algae, induced by excessive nutrients such as phosphorus (P) and nitrogen (N) that enter the lakes from the surrounding settlements. A typical example is the Sunway Lagoon (SL) lake located in the state of Selangor, Malaysia. With an area of 5 acres, the SL is relatively small compared to the surrounding catchment of 700 acres. Situated below mean sea level and without any natural outflow, SL is vulnerable to accumulation of nutrients and other pollutants flowing into the lake from the surrounding human settlements. After some 20 years in existence, the accumulation of nutrients in SL, particularly P and N, in the sediments is significant. The accumulated nutrients in the sediments are constantly released back into the water column to stimulate algae growth. Consequently, SL water quality has gradually deteriorated to the grade of Malaysia Class IIB, based upon the sample survey conducted on 24 to 25 June 2010. This deterioration of lake water quality is, however, common as virtually all tropical shallow lakes located within large human settlements will eventually become eutrophic. This unintentional "cultural" eutrophication in shallow tropical lakes is caused by accumulation of nutrients P and N in the sediment. It is important to minimize the accumulation of nutrients P and N in the sediment layers. This can best be achieved by flushing out water near the sediment layer, where the nutrients concentrations are much higher than those near the water surface. Reducing nutrients near the sediment will reduce the release of nutrients from the sediments back into the water column. High accumulation of nutrients in the sediments can induce high nonlinear interaction between sediments and water column, which can be facilitated via a low DO and a high pH environment. This undesirable water environment can be created by cyanobacteria activity that could promote the occurrence of highly undesirable and potentially irreversible hyper-eutrophic condition in SL in the future. Nutrients and heavy metals can potentially be released from the sediment layers into the water column if the aquatic conditions are suitable, for example if pH is above 8. Pore-water diffusion of nutrients will lead to potentially high concentration of P and N in the water column, especially near the bottom. For example, a P concentration of 20 mg/kg in the sediment could readily give rise to a P concentration of 1 mg/L in the water in immediate contact with the sediments if the pH is above 8, particularly at higher pH of 9 to 10. High pH could be facilitated and sustained by cyanobacteria activity promoted by high P concentration. This phenomenon of self-reinforcing feedback loop can induce a regime shift to hyper-eutrophication. High nutrient flushing rate will reduce the intensity of regime shift in SL and reduce the level of eutrophication. The objectives of this paper are: (1) to collect and collate environmental and limnology data relevant to eutrophication in SL, (2) to develop a robust mathematical model for regime shift in SL based upon the data compiled and (3) to propose effective eutrophication control measures for SL based upon the developed model that are low-cost and low-maintenance. Important regime shift thresholds critical to eutrophication control will be estimated for SL. This paper contributes to a development of methodology for eutrophication study and control that is applicable to all tropical shallow lakes worldwide.

2. Literature review on regime shift

There are ample signs of regime shift in SL to potentially undesirable hypereutrophic condition if the current trend is not reversed by means of effective control measures. A brief literature review on regime shift in ecosystems would provide useful scientific insights for this paper. The stability of ecosystems subject to external perturbations and internal feedback is a major topic in many fields in contemporary science [1], particularly for climate change studies. Increasing concern has been directed to the scenario known as catastrophic regime shift, where a relatively small change in the environmental conditions can lead to a sudden quantum jump from one state to another [2]. Catastrophic regime shifts are known to pose a serious threat to ecosystems, the control of which requires a reliable set of early warning indicators or thresholds [3]. Global climate change has the potential of inducing catastrophic regime shifts because of the complex nonlinear interactions across space and time involved in climate phenomena. Predicting the tipping point when a state begins to lose its stability and shifts to an alternative and undesirable state remains a daunting challenge. The mass conversion of heathland to grassland in north-western Europe during the past decades is a typical example of how tipping points threaten the biodiversity and ecosystem services delivered by biota [4]. Many large shallow lakes in China such as the Dianchi, Chaohu and Taihu, have already shifted to highly eutrophic conditions [5], that have vastly impaired the valuable ecosystem services provided. These lakes are plagued by persistent turbid water and annual toxic algae blooms, because of the excessive inputs of P and N from fertilizers over decades. Severe eutrophication in Lake Dianchi [5] in Kunming Yunnan (average depth of 4.4 m and surface area of 330 km^2) has impaired critical ecosystem services such as fresh water supply for 7 million people and ecotourism [6]. The ecosystem of Lake Chaohu (average depth of 3.1 m and surface area of 780 km^2) began to show a regime shift from the clear water state in early 1970s to the turbid state during early 1980s. Since the 1990s, the lake has been confronted with persistent signs of further regime shift to severe eutrophication [7]. At the low trophic level of 2.92, the trophic status of Lake Taihu is considered undesirable, due to the serious and persistent eutrophication in the lake [8, 9]. At these three large eutrophic lakes, the Chinese government has invested billions trying to prevent the occurrence of toxic algae blooms and to sustain portable water supplies for the surrounding cities, totalling more than 20 million people. The following three subsections provide a brief overview of the important concept of regime shift and its associated thresholds, which are highly relevant to this paper.

2.1. Positive feedbacks and bi-stability post deforestation

Availability of P in soils is a major factor limiting vegetation growth for natural ecosystems and agricultural systems [10]. P is not replenished biologically once it is lost from the soil [11]. Once removed from the soil, P becomes available for vegetation growth over very slow geologic time scales [12]. However, under conditions of limited P, soil microbes help in mitigating P losses and in enhancing P availability. Hence, the loss of microbial biomass due to deforestation increases P losses and decreases P availability. The interaction between vegetation growth, microbial abundance and P-cycling that drives the P-vegetation-microbial system was incorporated into a simulation model to study vegetation regime shift. Simulation results suggest that the systems most susceptible to regime shift are those that were previously deforested or those where the amount of P stored in the recalcitrant organic pool is low. A bi-stable equilibrium was observed after a significant loss of P exceeding a certain threshold. When deforestation occurred under limited P availability threshold, both the vegetation and the microbial biomass were not able to recover, and the system shifted to the low vegetation stable state for long duration [13].

2.2. Bi-stability of mangrove forests and freshwater plants

Remote sensing imagery in southern Florida confirms the sharp boundaries between freshwater hardwood hammock communities and halophytic communities such as mangroves. Competition among plant species [14, 15] and self-reinforcing feedback between coastal vegetation and vadose zone salinity are involved in maintaining this ecotone [16]. Along salinity gradients, this feedback creates a bi-stable equilibrium in which either halophytic habitat or freshwater plant communities dominate as alternative stable states. Literature shows how transpiration of mangroves and freshwater plants respond differently to vadose zone salinity, thus altering the salinity through feedback. Simulation models demonstrate how self-reinforcing feedback, together with physical template including salinity gradients and sunlight, control the bi-stability between halophytic and freshwater communities [17, 18]. Regions of bi-stability along gradients of salinity have the potential for large-scale regime shifts following large pulse disturbances such as hurricane, tidal surges in Florida, or tsunamis in other regions. The size of the region of bi-stability can be large for low-lying coastal habitats such as the Florida Everglades because of the high saline water table, which can extend several km inland due to salinity intrusion. This threat can be heightened potentially by climate change and sea level rise.

2.3. Bi-stability in reef oyster subject to sediment accumulation

Efforts over the past two decades to restore native oyster populations in the Chesapeake Bay have been extensive but largely ineffective [19]. However, recent restoration efforts and field experiments to revitalize reef oyster populations subject to sedimentation indicate that elevated reef height beyond certain thresholds can offset heavy sedimentation, promote oyster survival, encourage growth and improve disease resistance. This suggested the existence of alternative stable states for the ecosystem consisting of live oysters, reef height and sediment volume [20]. Bifurcation analysis performed on a system of three ordinary differential equations indicates that the initial reef heights dominate the outcomes of the equilibrium. This bifurcation analysis provided a theoretical framework for investigating strategy for restoration of degraded reef oyster populations. Alternative stable states can be triggered by environmental disturbances in various ecosystems, such as grasslands [21], kelp forests [22], coral reefs [23] and lakes [24]. For lakes, a system consisting of two ordinary differential equations (ODE) representing P and Chl-a [24] is adopted in this paper for analysing regime shift thresholds in eutrophic lake. Critical parameters determining the status of eutrophication are P loading into the lake (μg/L/day) and P flushing rate (day^{-1}) or its inverse, the P retention time (day).

3. Materials and methods

The SL lake has water inflow via surface runoff and groundwater but has no natural outflow, resulting in a persistent accumulation of P and N in the sediments. Currently, after each heavy rain, excess water in the lake is allowed to overflow a weir into the adjoining retention pond, to avoid flooding the vicinity. Some nutrients and pollutants are removed along with the surface water been drained; but the actual quantum of nutrients and pollutants been drained from SL is unknown. We propose to regularly pump off water at a deeper water level of 7.5 m to remove more nutrients. Quantum of water and key elements (P, N, BOD and Chl-a) removed will be recorded for regime shift simulation and analysis by means of equations (1) and (2), to be described in a later section.

3.1. Retention time R and flushing rate F

It is beneficial to remove nutrient-rich water located near the sediment layers to reduce the accumulation of nutrients in the sediments. For this regime shift analysis, we introduce the concept of retention time R (days) or its inverse i.e. the concept of flushing rate F (day^{-1}). Thus, we define $F = 1/R$. Consider a lake of volume $V = 10^6$ m^3, with a pollutant T of concentration C_T kg/m^3. Assume that the water is flushed out at the rate of $W = (10^3$ m^3 /day). Then the hydraulic retention time $R = V/W = 10^6$ m^3/ (10^3 m^3/day) = 10^3 days. The total mass M of T is $M = 10^6$ m^3 x C_T kg/m^3 = 10^6 C_T kg. Suppose now that the concentration of T near the bottom sediment is 10 times higher or 10 C_T kg/m^3. Further, suppose that we flush out the lake water near the bottom sediment (with concentration 10 C_T kg/m^3) at the same rate of $W = (10^3$ m^3/day). The mass of T flushed out per day is 10 C_T kg/m^3 x 10^3 m^3/day = 10^4 C_T kg/day. The retention time R_T of pollutant T is $R_T = 10^6$ C_T kg/10^4 C_T kg/day = 10^2 day; while the hydraulic retention time R is still 10^3 days. Therefore, the hydraulic flushing rate F is 1/10^3 day = 0.001 per day = 0.001 d^{-1}. On the other hand, the pollutant T flushing rate F_T is 1/10^2 day = 0.01 per day = 0.01d^{-1}. This means that $F_T = 10\ F$ or the pollutant T flushing rate F_T is 10 time higher than the hydraulic flushing rate F. It is therefore beneficial to flush out water near the sediment to remove as much nutrients as possible with the same hydraulic flushing rate F.

3.2. Fish tank experiment

A laboratory experiment in the staging facility in Sunway University was conducted to study the effect of fish respiration on dissolved oxygen (DO) level in water. For this purpose, 20 tilapias with different weights, ranging between 70 g to 335 g each, were selected and only one tilapia is placed into a tank for each experiment (figure 1). The dimension of tanks, water depth and water volume are presented in table 1. Figure 1 shows the fish tanks used in this experiment. The water in each tank is aerated overnight to saturation. In the next morning (at the start of the experiment), for each tank, the first readings of DO, temperature and water level are recorded, after which the aerator is turned off. Subsequently, DO level in the water is measured at regular intervals until the end of the day. The experiment is continued until 68 sets of data with various combinations of tank dimension and fish weight are obtained.

Figure 1. Fish tanks used in the experiment.

Table 1. Dimension, fish mass, and specific respiration rates of the five tanks in the experiment.

Tank	Length (m)	Width (m)	Water depth (m)	Water volume (L)	Fish mass (g)	Specific respiration rate (g O_2/kg fish/day)
1	0.780	0.340	0.177	46.94	85	3.5
2	0.590	0.295	0.156	27.15	95	2.1
3	0.800	0.340	0.185	50.32	155	3.0
4	0.750	0.370	0.226	62.72	200	3.7
5	0.750	0.370	0.269	74.65	335	2.4

3.3. Water and sediment quality sampling

A sampling survey was conducted on 24 to 25 June 2010 to define the water and sediment quality in SL. The water samples were collected at 10 spatially separated sampling locations at three water depths: surface (0.5 m), mid-depth (3.5 m) and lake bottom (7.5 m). Sediments were collected at 10 separate locations. Overall the water quality was classified as Malaysia Standard Class IIB. The surface water dissolved oxygen (DO) levels were generally below 5 mg/L, indicating a DO-stressed environment detrimental to aquatic lives including fish. In addition, a water quality and sediment quality survey was conducted on 23 January 2017. Water quality samples were collected at three depths of 0.5 m (surface), 3.5 m (mid-depth) and 7.5 m (bottom) below water surface. From December 2016 to May 2017, monthly water quality sampling was conducted at three depths (0.5, 3.5, 7.5 m below water surface) for six months. Water quality parameters sampled include temperature, pH, BOD, DO, ammoniacal nitrogen, nitrate nitrogen, phosphorus, chlorophyll-a, total coliform and *Escherichia coli*.

4. Water and sediment quality data

4.1 Water and sediment quality in 2010

Overall, DO varied between 1.26 mg/L to 4.74 mg/L from bottom to surface. Temperature was between 28 °C and 31 °C, and pH was neutral at 7. Concentration of chlorophyll-a (Chl-a) varied between 5 and 15 µg/L, suggesting a meso-eutrophic condition. With total kjeldahl nitrogen (TKN) concentration varying between 0.2 and 0.9 mg/L and total phosphorus (TP) between 0.1 and 0.4 mg/l, the water is considered as nutrients rich, capable of supporting vibrant Chl-a growth to a eutrophic level. The sediments were tested for 17 heavy metals at 10 locations. The heavy metal concentrations were generally low except for iron, aluminium and manganese. However, the concentrations for P and N are high, reflecting a eutrophic lake condition capable of supporting vibrant algae growth. Table 2 is a summary of concentrations of five key elements in the sediments sampled in 25 to 26 June 2010, indicating an undesirable water quality condition in the sediments.

Table 2. Concentrations of five key elements in the sediments 2010.

Component	Range of value	Level	Comment
TP (mg/kg)	22 to 81	High	Undesirable
TN (mg/kg)	68 to 246	High	Undesirable
TOC (% w/w)	0.4 to 1.5	High	Undesirable
Iron (mg/kg)	5225 to 13190	High	Undesirable
Aluminium (mg/kg)	8921 to 83890	High	Undesirable

4.2 Water and Sediment quality survey in January 2017

The results indicated that the overall water quality has deteriorated from Class IIB to Class III and the lake is eutrophic. Further, the N concentration in the sediments measuring 739 mg/kg (based upon an average of 3 sampling locations) has increased by 600 % as compared to that in 2010 (between 68 and 246 mg/kg based upon 10 sampling locations). Survey results for biochemical oxygen demand (BOD_5), DO and Chl-a at these 3 depths respectively indicated high and undesirable levels. The sediment quality measured on 23 January 2017 is summarized in table 3, indicating a water quality standard of Class III. Therefore, there is an urgent need to improve water quality in SL.

Table 3. Three sediment quality parameters sampled on 23 January 2017.

Component	Average value	Level	Comment
TP (mg/kg)	61.0	High	Undesirable
TN (mg/kg)	739	High	Undesirable
TOC (% w/w)	0.5	High	Undesirable

4.3 Monthly Water Quality Sampling Between 2016 and 2017

High levels of BOD, NH_3N and Chl-a recorded during the six months of sampling is shown in figure 2, indicating a eutrophic lake condition tipping gradually towards hyper-eutrophic. More data is available in [25]. A sustainable low-cost solution to control SL eutrophication is proposed in this paper. In general, an oligotrophic lake (not eutrophic) should have Chl-a level below 2 μg/L; while a meso-eutrophic lake should have Chl-a between 2 to 9 μg/L. Levels of Chl-a exceeding 9 μg/L is considered eutrophic and undesirable.

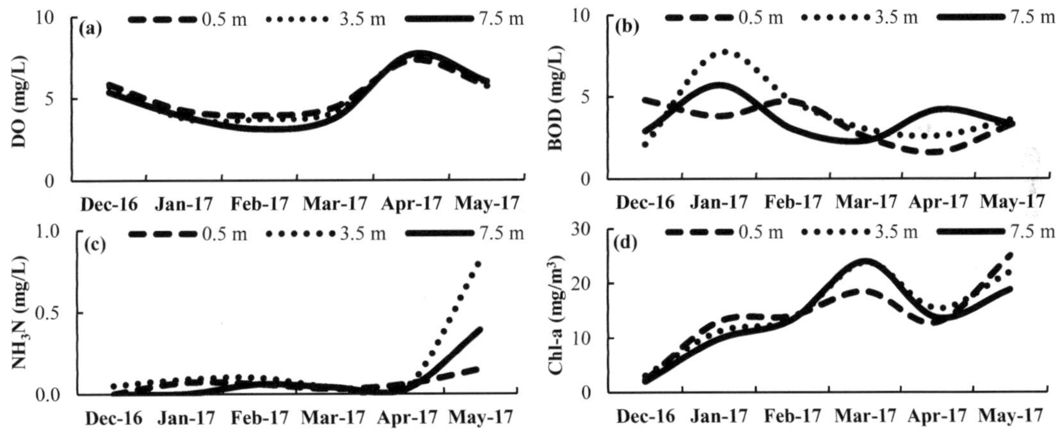

Figure 2. (a) DO, (b) BOD, (c) NH_3N and (d) Chl-a sampled at three depths (0.5, 3.5 and 7.5 m below water surface) from Dec 2016 to May 2017.

5. Simulation results

To control SL eutrophication, a simple, sustainable and cost-effective method is proposed. Lake water at a depth of 7.5 m will be removed at daily frequency to flush out P-rich lake water [25]. The system of two ODE representing the evolution of Chl-a (equation (1)) and P (equation (2)) in SL is given in equations (1) and (2), with the definitions, units, and values of process parameters shown in table 4. Further details are available elsewhere [26]. The P-flushing rate denoted by F_T (day^{-1}) is represented by h in equation (2). Three flushing scenarios are simulated: (a) h = 0 day^{-1}, (b) h = 1/(200 days) = 0.005 day^{-1} and (c) h = 1/(100 days) = 0.01 day^{-1}, representing zero flushing, flushing with 200-days retention and flushing with 100-days retention respectively. The corresponding simulation results are demonstrated in figure 3.

$$\frac{dA}{dt} = bPA - \left(g + \frac{s}{z_e} + h \right) A \tag{1}$$

$$\frac{dP}{dt} = l + r + egA - bAP - hP \tag{2}$$

Table 4. Definition, units and value of variables and parameters.

Symbol	Definition	Unit	Value
b	Algal growth rate per unit	L/μg P/d	0.900
g	Zooplankton grazing rate	d^{-1}	0.030
s	Algal sinking rate	m/d	0.085
h	Flushing rate	d^{-1}	(a) 0.0, (b) 0.005, (c) 0.01
e	*P* excretion rate associated with grazing	μg P/μg chl	0.650
l	External *P* loading rate	μg P/L/d	0.01 – 0.13
r	Internal *P* sediment recycling rate	μg P/L/d	Computed
z_e	Thickness of epilimnion	m	7.000

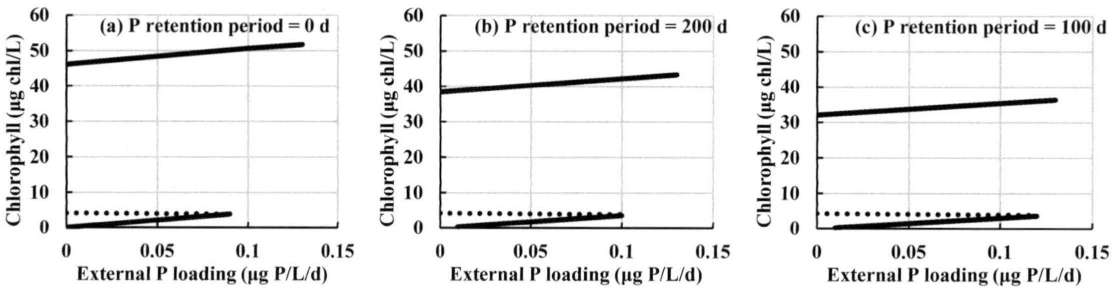

Figure 3. Simulated results for three scenarios of retention period: (a) No flow ($h = 0.0$ d^{-1}), (b) 200 days ($h = 0.005$ d^{-1}) and (c) 100 days ($h = 0.01$ d^{-1}).

With zero flushing, regime shift tipping point is reached at an external P loading rate of 0.8 μg/L/d, and the Chl-a concentration may peak at 50 μg/L, as may be seen from figure 3(a). A P flushing rate of 0.005 day^{-1}, corresponding to a retention time of 200 days, can extend the regime shift tipping point to external P loading of 1.0 μg/L/d, and the Chl-a concentration may peak at 40 μg/L as illustrated in figure 3(b). A P flushing rate of 0.01 day^{-1}, corresponding to a retention time of 100 days, can extend the regime shift tipping point to external P loading of 1.2 μg/L/d, and the Chl-a concentration may peak at 35 μg/L, as may be seen from figure 3(c). Given that the lake is shallow, and the temperature is high all year round, SL is vulnerable to eutrophication [24].

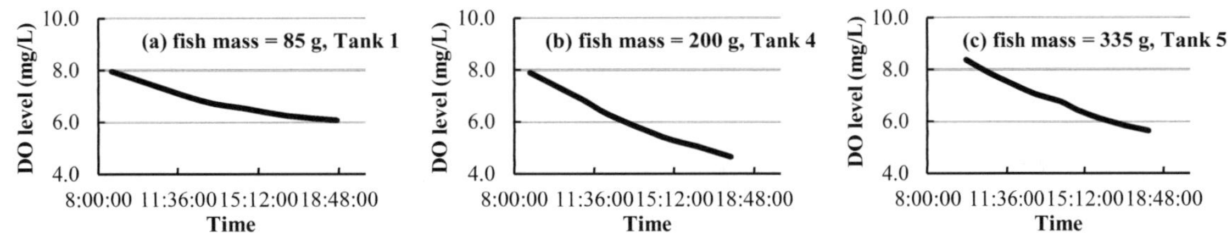

Figure 4. Dissolved oxygen in Tank (a) 1, (b) 4 and (c) 5.

Hence, in addition to increasing P flushing, other additional measures should be considered, including reduction of fish density. For this purpose, a series of lab experiments as described in section 3.2 was conducted to measure fish specific respiration rates under lab conditions. Figure 4 demonstrates the decline in DO due to fish respiration in three of the fish tanks. The dimension, fish mass, and specific respiration rates in the five tanks are given in table 1, indicating specific respiration rate of between 2.1 to 3.7 g O_2/kg fish/day for sedentary fish at 23 °C that are inactive most of the time. In SL where fish compete for resources in an environment of higher temperature of around 29 °C, the specific respiration rate should be higher. Numerical simulations were performed for a fish pond with depth of 0.6 m, reaeration rate of 2 d^{-1}, and specific respiration rate of 5 g O_2/kg fish/day, fish feeding rate of 2 % per day wet weight and fish density of 0.8 kg/m^3. The simulation results indicated fish will be killed because of low DO, without an aerator [27]. It is therefore important to reduce fish density in SL to 0.1 kg/m^3 or less to ensure adequate DO. With a depth of 8 m, SL reaeration rate will be less than 2 d^{-1}, limiting oxygen replenishment from atmosphere.

6. Discussion and conclusion

Although the causes of lake eutrophication regime shift have been broadly understood, the theoretical foundation of eutrophication is still lacking. Hence, eutrophication remains a daunting environmental problem that is difficult to resolve. With high annual temperature and abundant sunlight in the tropics, eutrophication is accelerated by large and persistent influx of nutrients such as P generated by anthropogenic activities. A high concentration of these nutrients in the lake stimulates algae to grow wildly, leading to depletion of DO in the lake water. The accumulation of P in the sediment and the release of P from sediment back into the water column are dominant processes driving regime shift. The persistence of eutrophication is dominated by important ecological thresholds and the existence of multiple regimes within a lake. The knowledge on regime shift thresholds associated with eutrophication is a prerequisite in lake restoration efforts to improve the lake ecosystem. The analysis performed in this paper provides the fundamental knowledge and insights on the regime shift and its thresholds for SL and offers sustainable solutions for controlling eutrophication. It is concluded that a sustainable and cost-effective rehabilitation program for SL is a frequent removal of nutrient rich lake water from a depth of 7.5 m. This daily flushing of P rich water will delay the tipping point of lake regime shift from 0.8 μg/L/d to 1.2 μg/L/d and significantly reduce P and Chl-a level. Reduction of fish density to 0.1 kg/m^3 or less may also be beneficial in improving overall water quality and increasing DO levels. Increased DO may inhibit release of P from sediment to water column. The methodology developed in this paper may be applicable to other tropical shallow lakes for the analysis and control of eutrophication.

7. References

[1] Müller F, Burkhard B and Kroll F 2009 *Resilience, Integrity and Ecosystem Dynamics: Bridging Ecosystem Theory and Management* ed J C Otto and R Dikau (Berlin: Springer) pp 221–42

[2] Scheffer M, Carpenter S R, Lenton T M, Bascompte J, Brock W, Dakos V, Van De Koppel J, Van De Leemput L A, Levin S A, Nes E H V et al. 2012 Anticipating critical transition *Science* **338** pp 344–8

[3] Weissmann H and Shnerb N M 2016 Predicting catastrophic shifts *J. Theor. Biol* **397** pp 128–34

[4] van Voorn G A K, Kooi B W and Bregt A K 2016 Over-shading is critical for inducing a regime shift from heathland to grassland under nitrogen enrichment *Ecol. Complex.* **27** pp 74–83

[5] Cao X, Wang Y, He J, Luo X and Zhang Z 2016 Phosphorus mobility among sediments, water and cyanobacteria enhanced by cyanobacteria blooms in eutrophic Lake Dianchi *Environ. Pollut.* **219** pp 580–7

[6] Huang C, Wang X, Yang H, Li Y, Wang Y, Chen X and Xu L 2014 Satellite data 9-15 regarding the eutrophication response to human activities in the plateau lake Dianchi in China from 1974 to 2009 *Sci. Total Environ.* **485–486** pp 1–11

[7] Kong X, Dong L, He W, Wang Q, Mooiji W M and Xu F 2015 Estimation of the long-term nutrient budget and thresholds of regime shift for a large shallow lake in China *Ecol. Indic.* **52** pp 231–44

[8] Wang F E, Tian P, Yu J, Lao G M and Shi T C 2011 Variations in pollutant fluxes of rivers surrounding Taihu Lake in Zhejiang Province in 2008 *Phys. Chem. Earth Parts A/B/C* **36** pp 366–71

[9] Bai X L, Ding S M, Fan C X, Liu T, Shi D and Zhang L 2009 Organic phosphorous species in surface sediments of a large, shallow, eutrophic lake, Lake Taihu, China *Environ. Pollut.* **157** pp 2507–13

[10] Wardle D A, Walker L R and Bardgett R D 2004 Ecosystem properties and forest decline in contrasting longterm chronosequences *Science* **305** pp 509–13

[11] Vitousek P M, Porder S, Holton B Z and Chadwick O A 2010 Terrestrial phosphorous limitation: mechanisms, implications and nitrogen-phosphorous interactions *Ecol. Appl.* **20** pp 5–15

[12] Chadwick O A, Derry L A, Vitousek P M, Heubert B J and Hedin L O 1999 Changing sources of nutrients during four million years of ecosystem development *Nature* **397** pp 491–6

[13] Runyan C W and D'Odorico P 2013 Positive feedbacks and bistability associated with phosphorus-vegetation-microbial interactions *Adv. Water Resour.* **52** pp 151–64

[14] McKee K L 2011 Biophysical controls on accretion and elevation change in Caribbean mangrove ecosystems *Estuar. Coast. Shelf Sci.* **91** pp 475–83

[15] Medina E, Cuevas E and Lugo A E 2010 Nutrient relations of dwarf Rhizophora mangle L mangroves on peat in eastern Puerto Rico *Plant Ecol.* **207** pp 13–24

[16] Teh S Y, DeAngelis D L, Sternberg L D L, Miralles-Wilhelm, F R, Smith T J and Koh H L 2008 A simulation model for projecting changes in salinity concentrations and species dominance in the coastal margin habitats of the Everglades *Ecol. Model.* **213** pp 245–56

[17] Jiang J, Fuller D O, Teh S Y, Zhai L, Koh H L, DeAngelis D L and Sternberg L D S L 2015 Bistability of mangrove forests and competition with freshwater plants *Agr. Forest Meteorol.* **213** pp 283–90

[18] Teh S Y, Koh H L, DeAngelis D L, Voss C I and Sternberg L D S L 2019 Modeling $\delta^{18}O$ as an early indicator of regime shift arising from salinity stress in coastal vegetation *Hydrogeol. J.* pp 1–20

[19] Ruesink J, Lenihan H, Trimble A, Heiman K, Micheli F, Byers J and Kay M 2005 Introduction of non-native oysters: ecosystem effects and restoration implications *Annu. Rev. Ecol. Evol. S.* **36** pp 643–89

[20] Jordan-Cooley W C, Lipcius R N, Shaw L B, Shen J and Shi J 2011 Bistability in a differential equation model of oyster reef height and sediment accumulation *J. Theor. Biol.* **289** pp 1–11

[21] Rietkerk M and Van de Koppel J 1997 Alternate stable states and threshold effects in semi-arid grazing systems *Oikos* **79** pp 69–76
[22] Konar B and Estes J 2003 The stability of boundary regions between kelp beds and deforested areas *Ecology* **84** pp 174–85
[23] Hughes T 1994 Catastrophes, phase shifts, and large-scale degradation of a Caribbean coral reef *Science* **265** pp 1547–51
[24] Carpenter S, Ludwig D and Brock W 1999 Management of eutrophication for lakes subject to potentially irreversible change *Ecol. Appl.* **9** pp 751–71
[25] Koh H L 2017 *Sunway Lagoon Rehabilitation Study: Data, Simulation and Synthesis* (Selangor: Sunway Lagoon Theme Park)
[26] Koh H L, Tan W K and Teh S Y 2018 Regime shift analysis and numerical simulation for effective ecosystem management *Int. J. Environ. Sci. Dev.* **9** pp 192–9
[27] Koh H L, Teh S Y, Lee E, Tan W K, Sagathevan K A and Low A 2018 Derivation of optimal fish stocking density via simulation of water quality model E2Algae *AIP Conf. Proc.* **1974** 020042

Acknowledgements
This work was supported by Kementerian Pendidikan Malaysia FRGS Grant awarded to TSY and the Sunway University Internal Research Grant Scheme INT-2018-SMS-05 awarded to TWK. Financial, administrative and technical support provided by Sunway Lagoon Theme Park and the conducive research environment provided by JSC to KHL is appreciated.

Land use change and fragmentation in the protected areas of peninsular Malaysia: The cases of Klang Gate and Sungai Dusun wildlife reserves

W H Lee[1*], S A Abdullah[1] and N H Rafaai[1]

[1]Institute for Environment and Development (LESTARI),
Universiti Kebangsaan Malaysia
43600 UKM Bangi, Selangor,
Malaysia.

**Corresponding author: nicole_kwy@yahoo.com*

Abstract. Ongoing economic and demographic growth in developing countries has exponentially increased the need for anthropogenic land use. The expansion of human populations into pristine environments has publicized the fierce competition between environmental conservation and anthropogenic resource demands. Because protected areas (PAs) are increasingly imperilled by human land use, particularly in the tropical regions, PAs should be regularly monitored and assessed for land use modifications in order to determine the impact on long term sustainability. In this work, land use change assessments and forest fragmentation analysis were conducted for the period between 1988 and 2012 at two PAs in peninsular Malaysia — Klang Gate and Sungai Dusun wildlife reserve. The findings demonstrated that Klang Gate lost 29.9% of its forest cover due to surrounding development and commercial agriculture; while commercial agriculture in Sungai Dusun caused only a 5% forest loss. Furthermore, Klang Gate also showed a higher degree of forest fragmentation when compared to Sungai Dusun. The fact that Klang Gate experienced ~6x more habitat loss than Sungai Dusun implies that area underwent unsustainable development. Thus, more in depth evaluation regarding the effectiveness of conservation measures within PAs of developing countries is crucial to identifying gaps in management and conservation policy.

1. Introduction

Over the past two decades, there has been a significant increase in the number of PAs worldwide, as scientists and governments attempt to protect their country and continent's remaining biodiversity, before it's completely destroyed by irresponsible development and other human activities [1,2]. Due to increasing conservation measures, to date, PAs account for ~15% of global terrestrial and inland water environments. However, additional investment and support is required to reach the 2020 target of 17% coverage, outlined in the Convention on Biological Diversity (CBD) [3]. Nevertheless, despite these conservation efforts, rich natural heritage in Asia, and specifically Southeast Asia, is still imperilled by extensive anthropogenic threats, primarily habitat loss and environmental degradation [4,5].

The habitat loss and environmental degradation are directly correlated to anthropogenic land use – particularly forest fragmentation and natural vegetation destruction – as they produce significant landscape changes within the PAs [6]. In essence, the needs of a continuously expanding population and the incentives to promote economic growth are completely at odds with the goals of global PA network expansion; and intense competition has developed between environmental conservation and anthropogenic resource demands [7]. This conflict is especially prevalent in rapidly developing countries, like Peninsular Malaysia.

Peninsular Malaysia is among the world's most biologically diverse regions and is one of the fastest developing countries. While Peninsular Malaysia harbours many tropical species, evidence suggests that tropical forests in this region are undergoing significant land use and landscape alterations, putting many species at risk for extinction due to detrimental anthropogenic activity (e.g. [8,9]). For this reason, the 2016 – 2025 National Policy on Biological Diversity is focused on addressing the direct and indirect pressures affecting biodiversity in order to safeguard the ecosystem and species diversity [10].

To date, studies that analyse and evaluate land use and landscape alteration in the PAs of developing countries are still scarce, which prevents effective planning and management of these sites and limits the ability to effectively strategize an integrated and comprehensive approach to land use planning. Thus, whether or not PAs in developing regions will be mildly, moderately, or severely affected by residential, agricultural, and industrial expansion, remains unknown. As such, the objective of this study is to analyse the land use and forest fragmentation of two PAs in Selangor state, Peninsular Malaysia, in order to evaluate the effects of increasing anthropogenic influence between the years 1988 and 2012. Two wildlife reserves were selected as case studies— Klang Gate wildlife reserves (hereafter referred to as Klang Gate) and Sungai Dusun wildlife reserves (hereafter referred to as Sungai Dusun). The terms *wildlife reserve* and *protected area* are used interchangeably throughout this work, as wildlife reserve is a type of PA that is established primarily for wildlife protection and biodiversity conservation.

2. Materials and methods

2.1. Study sites

The two selected wildlife reserves are located in Selangor State (Table 1) — the most developed and populous state in Peninsular Malaysia (figure 1). It also yields the highest gross domestic product per capita [11]. This state has a highly diversified economy; composed of agriculture, manufacturing, industry, commerce, and tourism. Therefore, extensive study in this state is essential to enhancing the land use planning and conservation effort.

Table 1. Brief details for the wildlife reserves that selected as case studies.

Wildlife Protected Areas	IUCN Category	Land Area Management	Area (ha)	Location	Elevation (meter above sea level)
Klang Gate	VI[c]	DWNP[a]	1348	3°12′-3°16′N & 101°43′-101°46′E	130 – 400
Sungai Dusun	Ib[b]	DWNP	5113	3°39′-3°42′N & 101°20′-101°29′E	25 – 253

[a]Department of Wildlife and National Parks.
[b]IUCN category VI – Wilderness area.
[c]IUCN category Ib – Protected area with sustainable use of natural resources.

Klang Gate was selected as a study site because it is a PA located adjacent to an urbanizing area. As such, it makes an ideal place to evaluate: 1) the effects of development and urbanization pressures on a conservation site, and 2) the effectiveness of current conservation measures in protecting the PAs against development and urbanization pressures. Additionally, Klang Gate supports a quartz ridge more than 14 km long and 200 m wide, which is considered one of the most unique geologic formations in Malaysia. This ridge also hosts five endemic plant species – *Aleisanthia rupestris*, *Borreria pilulifera*, *Eulalia milsumii*, *Henckelia primulina*, and *Ilex praetermissa* [12,13] and an endangered animal called the Sumatran serow (*Capricornis sumatraensis*) [14]. Sungai Dusun was selected as the second study site because it is also a strictly PA, but is surrounded only by oil palm plantations and is likely to be affected exclusively by agricultural development. Furthermore, endangered species such as the Malayan tapir (*Tapirus indicus*) and the white-handed gibbon (*Hylobates lar*) also exist within the reserve. Hence,

Klang Gate and Sungai Dusun are both critical refuges for biodiversity surrounded by different levels of encroaching development. As such, they represent an excellent opportunity to evaluate how well these PAs resist the surrounding land use change pressure.

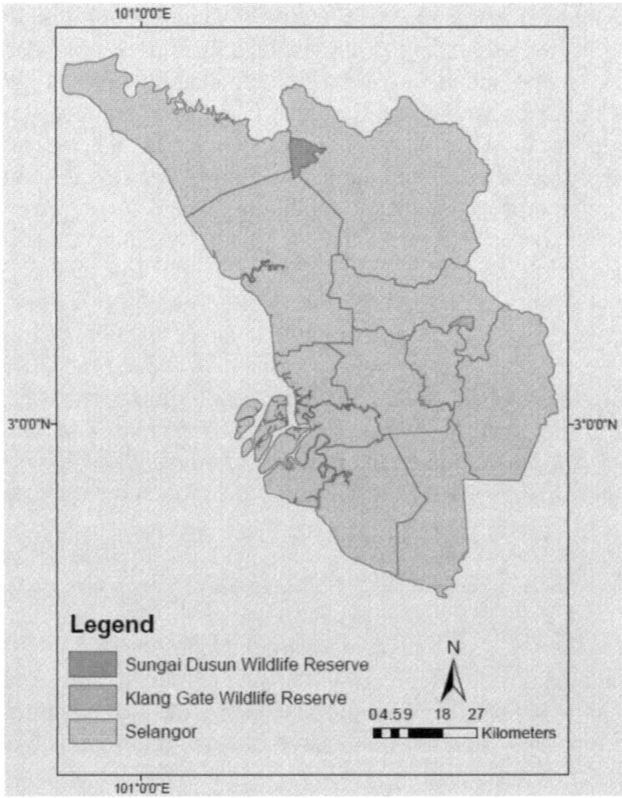

Figure 1. Location of two study sites in Selangor state.

2.2. Data source and processing of images

A 1988 land use map of Klang Gate and Sungai Dusun was obtained from the Institute for Environment and Development (LESTARI), Universiti Kebangsaan Malaysia (refer to [15,16]). The map was compiled using 30m Landsat Thematic Mapper (TM) images. The 2012 land use maps of the reserves were obtained from the Malaysian Remote Sensing Agency (MRSA). The maps were based on 10m resolution SPOT 5 images that were processed using ERDAS Imagine 9.1 software.

The SPOT 5 satellite images were geo-corrected to align with the GCS Kertau coordinate system. After creating a mosaic out of the individual images, boundaries were delineated to identify the area that defined the PAs. Bands of spectral data were combined to generated band combinations that enhance interpretation and differentiation of various land uses of interest. A false colour composite of band combination 4-3-2 was applied to the images. Finally, the images were classified using supervised classification that was primarily controlled by the analyst [17,18]. The land use classifications for the 1988 Landsat TM imagery included: built-up area, cleared land, commercial agriculture, paddy and other agriculture, water body, and forest. The same classification system was used for the 2012 SPOT 5 imagery. Classification accuracy was assessed by calculating the overall accuracy, *po*, and the kappa coefficient, *k*. The overall classification accuracies for Klang Gate and Sungai Dusun were 88.28% and 83.59%, respectively, and *k* were 0.84 and 0.80, respectively.

IOP Conf. Series: Earth and Environmental Science **380** (2019) 012007 doi:10.1088/1755-1315/380/1/012007

2.3. Data analysis

Fragmentation analysis was conducted on both study sites. However, to account for differences in the spatial resolution of the 1988 (30m resolution) and 2012 (10m resolution) maps, 2012 data was resampled using the nearest neighbour algorithm to a 30m resolution.

Fragmentation was analysed using GUIDOS, a standalone freeware toolbox [19]. Details of this fragmentation analysis method are reported in [20]. The raster images for each study site were analysed based on seven geometric categories [21,22]: core, islet, loop, bridge, perforation, edge, and branch (refer to [20] for category definitions). Once the category percentage was finalized, equation (1) was used to quantify the degree of fragmentation:

$$F = 100 - [C - (I + P + E + B + R + L) \quad (1)$$

where F is the percentage of total fragmentation; C represents the percentage of the core area; *I, P, E, B, R, and L* represent the percentage of the islet area, perforated area, edge area, branch area, bridge area, and loop area, respectively. An *F* value of 100 indicates the highest fragmentation, whereas 0 implies that the protected area is fragmentation free.

3. Results and discussion

3.1 Land use change at Klang Gate from 1988 – 2012

Land use categories identified at Klang Gate in 1988 consisted of forest, water body, built-up area, and cleared land. Forest dominated the land use in Klang Gate for both years, with 87.21% in 1988 and 57.32% in 2012 (figure 2, table 2). In 1988, forest coverage was followed by water body (11.67%); and the remaining land uses - cleared land and built-up area - represented less than 2% of the total land use (figure 2a). However, by 2012, built-up area became the second highest land use (15.61%) followed by commercial agriculture (13.31%) and the water body (12.79%).

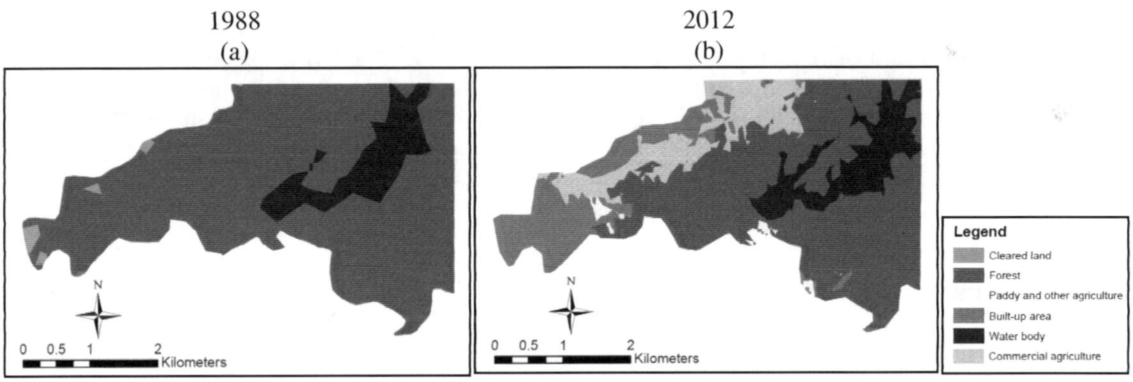

Figure 2. Land use map for Klang Gate at 1988 (a) and 2012 (b).

Table 2. Land use distribution of Klang Gate and Sungai Dusun from 1988 – 2012.

Land use types	Klang Gate				Sungai Dusun			
	1988 (%)	2012 (%)	% Changes	Annual change rate	1988 (%)	2012 (%)	% Changes	Annual change rate
Forest	87.22	57.32	– (34.28)	– (1.25)	98.04	93.02	– (5.12)	– (0.21)
Water body	11.67	12.79	+ (9.60)	+ (0.05)	0	0	0	0
Commercial agriculture	0	13.31	+ (100)	+ (0.55)	1.96	6.80	+ (247.45)	+ (0.2)
Paddy and other agriculture	0	0.97	+ (100)	+ (0.04)	0	0	0	0
Built-up area	0.07	15.61	+ (22,200)	+ (0.65)	0	0.18	+ (100)	+ (0.01)
Cleared land	1.04	0	– (100)	– (0.04)	0	0	0	0

Land use changes between 1988 and 2012 show that forest cover and cleared land reduced throughout this period, in which forest cover declined by a remarkably 29.9%, while cleared land only slightly declined by 1.04% (figures 2b and 3). Both forest cover and cleared land show negative trends (–34.28% and –100%, respectively). Apart from these two land use categories, the coverage of other land uses increased, demonstrating a positive trend over the 24–year period (Table 2). Built-up area increased 15.54%, depicting the largest increase among all the categories and it shows a positive change (22,200%) (Table 2). The built-up area's annual rate of change was 0.65% while the remaining land uses showed less than 0.1% annual rate of change (Table 2).

3.2 Land use change at Sungai Dusun from 1988 – 2012

Land use categories identified at Sungai Dusun in 1988 included only forest and commercial agriculture. This wildlife reserve was predominantly covered by forest (98.04%), whereas commercial agriculture occupied only 1.96% of the total land use (figure 4a, table 2). By 2012, built-up area was introduced to Sungai Dusun and contributed to 0.18% of the land use (Figure 4b). Forest cover in this reserve presented a negative pattern of change (–5.12%) and dropped to 93.02% during the study period (Figure 4b, Table 2). While forest cover dropped 5.02%, commercial agriculture and built-up area increased to 4.84% and 0.18%, respectively (Figure 3). Therefore, commercial agriculture and built-up area show a positive change pattern (+247.45% and +100%) (Table 2). Since the magnitude of change for each land use is generally low, their annual rate of change was less than 0.3%.

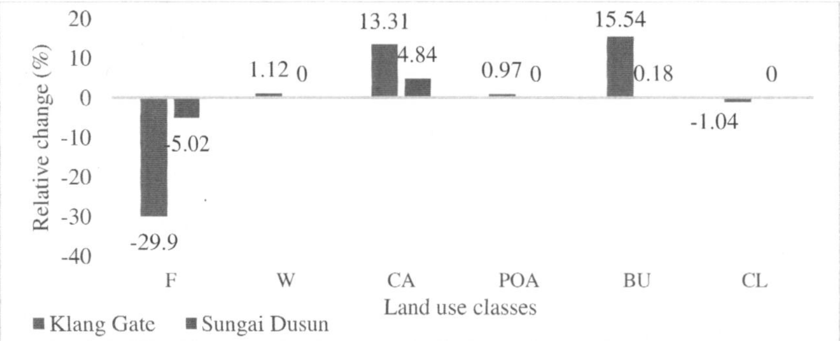

Figure 3. Relative change of percentage in land use for Klang Gate and Sungai Dusun. F – forest; W – water; CA – commercial agriculture; POA – paddy and other agriculture; BU – built-up area; CL – cleared land.

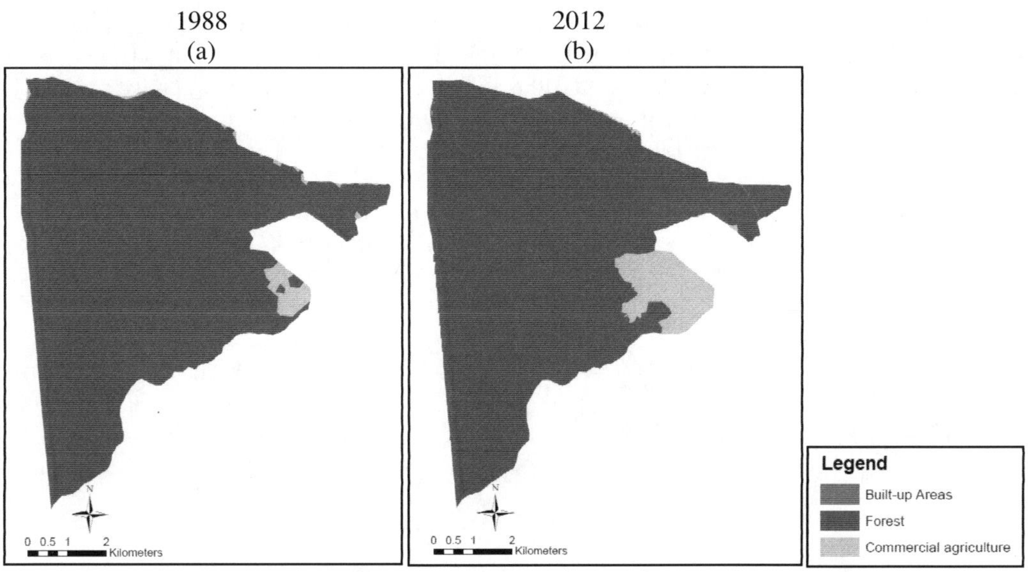

Figure 4. Land use map for Sungai Dusun at 1988 (a) and 2012 (b).

3.3 Overall land use pattern at Klang Gate and Sungai Dusun from 1988 – 2012

In Klang Gate, forest cover loss and agricultural land growth were both observed throughout the temporal study period. These results imply that socio-economic growth has adversely impacted this reserve. The land use and land cover changes demonstrate that the dynamic and complex interactions between structural and behavioural factors associated with social, technological, and infrastructure capacity demands; and the biophysical attributes of the area can and have become so unbalanced that they eventually affect the environmental health of the ecosystem [23,24]. The increase in population growth exerts pressure on limited natural resources and subsequently contributes to various negative land use changes in order to meet anthropogenic demand [23,25].

This scenario was observed at Klang Gate, where agricultural land encroached closer to the PA core and was followed by built-up area at the frontiers; resulting in ~ 30% forest loss within a 24–year period. Furthermore, land use policy is one of the most prominent factors that influence land use within wildlife reserves. In the early years after national independence, land development policies in Peninsular Malaysia strongly favoured agricultural expansion (i.e., rubber and oil palm plantation); that trend changed in the 1980s to supporting development of manufacturing. As one of the most rapidly developing states, Selangor state exemplifies the national policies. Urban development at Kuala Lumpur city centre progressed rapidly and intensely; and facilities, residential areas, and industrial sectors have made it a centre of development [24]. Klang Gate is located in close proximity to Kuala Lumpur city centre; therefore, the human impact and encroachment are stronger compared to the PAs located far from developing areas.

Sungai Dusun shows ~ 5% forest loss over the past two decades, primarily due to encroachment of oil palm plantations in the southeast region. Sungai Dusun is bounded by several Federal Land Development Authority (FELDA) projects that converted forest to oil palm plantations. Examples include, but are not limited to, Lima Belas Estate, FELDA Gedangsa and FELDA Soeharto to the north, and FELDA Sungai Tenggi to the east. This reserve is located far away from high population density. Moreover, the biophysical conditions, i.e., soil, geography, drainage condition, proximity to water resources, and suitability to be converted to agriculture or settlement determines appropriate land use of this reserve [24,26]. Sungai Dusun is characterised by the presence of lowland dipterocarp and fresh water peat swamp forest with an elevation distribution between 25 and 253 m.a.s.l. Hence, topography of this reserve is suitable for commercial plantation as opposed to high population settlement.

3.4. Landscape pattern change of Klang Gate and Sungai Dusun from 1988 – 2012

Landscape pattern changes in the wildlife reserves were assessed using fragmentation analysis. Fragmentation in Klang Gate increased from 7.06% in 1988 to 35.39% in 2012 (Figure 5). Fragmentation in Sungai Dusun was 9.41% in 1988 - slightly higher than Klang Gate – but increased only to 14.39% by 2012. Hence, 2012 fragmentation of Klang Gate was significantly higher than that of Sungai Dusun.

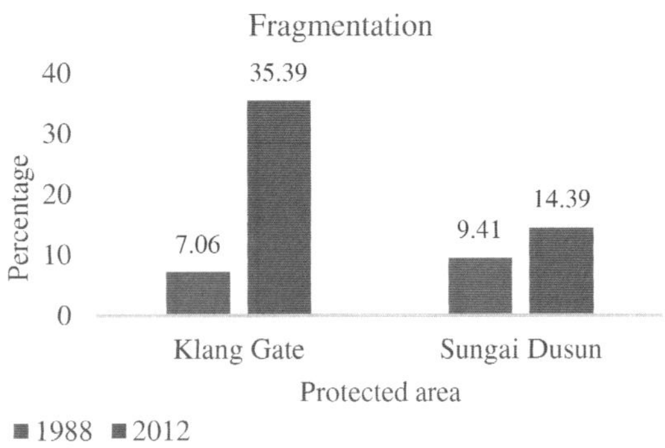

Figure 5. Fragmentation for Klang Gate and Sungai Dusun at 1988 and 2012.

Inevitably, fragmentation significantly increases the edge width of the PAs and creates negative edge impacts (e.g. fluxes of radiation, wind and water and lead to microclimate change) [27,28]. Subsequently, fragmentation can lead to isolation and habitat loss, which adversely effects biodiversity [29], such as local extinction to species that depend on large, intact forest. Furthermore, when humans fragment forests and take over land for anthropogenic use in the PAs, other anthropogenic threats emerge, such as poaching, excessive resources extraction, human-induced forest fires, and introduction of invasive species. As such, this assessment of land use change using remote sensing data is likely an underestimate of the actual loss in ecological integrity; as this method fails to account for human impacts that can't be evaluated on a landscape scale.

4. Conclusion

In this study, land use changes were reported within Klang Gate and Sungai Dusun. Agricultural expansion and residential growth that resulted from the urbanizing processes at the nearest city— Kuala Lumpur, were predominantly increasing human impacts and fragmentation within these PAs. These unsustainable anthropogenic influence in the highly developing region could cause habitat loss, ecological integrity degradation, conservation capacity decline, wildlife movement inhibition, and eventually leads to biodiversity loss in the PAs. Since wildlife movement and flow are not confined to any physical boundaries of a PA, more in-depth evaluation regarding the land use at the PA surroundings is crucial to quantify their integrity. Furthermore, appropriate collaboration between policy makers and other stakeholders (e.g., park managers, scientists, land tenures, local communities and NGO) is necessary to halt the land conversion in PAs and initiate forest restoration in the PAs that were degraded by human land use.

5. References

[1] Locke H and Dearden P 2005 Rethinking protected area categories and the new paradigm *Environ. Conserv.* **32** 1–10

[2] Saviano M, Di Nauta P, Montella M and Sciarelli F 2018 Managing protected areas as cultural landscapes: The case of the Alta Murgia National Park in Italy *Land Use Policy* **76** 290–9

[3] UNEP-WCMC, IUCN 2016 Protected Planet Report 2016 (Cambridge UK and Gland, Switzerland: UNEP-WCMC and IUCN) pp 30

[4] Juffe-Bignoli D, Bhatt S, Park S, Eassom A, Belle E M S, Murti R, Buyck C, Raza Rizvi A, Rao M, Lewis E, MacSharry B and Kingston, N 2014 *Asia Protected Planet Report 2014* (Cambridge: UNEP-WCMC)

[5] Wilkinson C L, Yeo D C J, Tan H, Fikri A H and Ewers R M 2018 Land-use change is associated with a significant loss of freshwater fish species and functional richness in Sabah, Malaysia *Biol. Conserv.* **222** 164–171

[6] Verburg P H, Overmars K P, Huigen M G A, Groot W T De and Veldkamp A 2006 Analysis of the effects of land use change on protected areas in the Philippines. *Appl. Geogr.* **26** 153–173

[7] Gross D, Dubois G, Pekel J F, Mayaux P, Holmgren M, Prins H T, Rondinini C and Boitani L 2013 Monitoring land cover changes in African protected areas in the 21st century *Ecol. Inform.* **14** 31–37

[8] Abdullah S A, Nor S M and Yusof A M 2014 Understanding development trends and landscape changes of protected areas in peninsular Malaysia: a much needed component of sustainable conservation planning (eds) Nakagoshi N, Mabuhay J A, *Designing Low Carbon Societies in Landscapes, Ecological Research Monographs* (Japan: Springer) pp 205–222

[9] Aiken S R 1994 Peninsular Malaysia's protected areas' coverage, 1903-92: creation, rescission, excision, and intrusion *Environ. Conserv* **21**: 49–56

[10] Ministry of Natural Resources and Environment (NRE) 2016 *National Policy on Biological Diversity 2016 – 2025* (Putrajaya: NRE)

[11] Abdullah S A and Nakagoshi N 2006 Changes in landscape spatial pattern in the highly developing state of Selangor, peninsular Malaysia *Landsc. Urban Plan.* **77** 263–275

[12] Reid J A 1951 Klang Gates and Bukit Takun: Reflections of an Amateur Botanist. *Malayan Nature Journal* **5** 109–123

[13] Kiew R 1982 The Klang Gates ridge *Malay Nat* **36** 22–28

[14] Perumal B 1992 *Management Survey of Klang Gates Wildlife Sanctuary* (Petaling Jaya: WWF Malaysia)

[15] Reza M I H 2014 Measuring forest fragmentation in the protected area system of a rapidly developing Southeast Asian tropical region Science Postprint **1** 1–7

[16] Reza M I H, Abdullah S A, Nor S B M and Ismail M H 2013 Integrating GIS and expert judgment in a multi-criteria analysis to map and develop a habitat suitability index: A case study of large mammals on the Malayan Peninsula *Ecol. Indic.* **34** 149–158

[17] Bailey K M, Mccleery R A, Binford M W, Zweig C, Bailey K M, Mccleery R A, Binford M W, et al 2016 Land-cover change within and around protected areas in a biodiversity hotspot *J. Land Use Sci.* **11** 154–176

[18] Eastman J R 2001 *Idrisi 32 release 2 guide to GIS and image processing vol 1* (Worcester: Clark University)

[19] European Commission 2019 GuidosToolbox (GTB). https://forest.jrc.ec.europa.eu/en/activities/lpa/gtb/ (online 20 December 2017).

[20] Clay E, Moreno-Sanchez R, Torres-Rojo J M and Moreno-Sanchez F 2016 National assessment of the fragmentation levels and fragmentation-class transitions of the forests in Mexico for 2002, 2008 and 2013 *Forests* **7** 6–8

[21] Soille P and Vogt P 2009 Morphological segmentation of binary patterns *Pattern Recognit. Lett.* **30** 456–459

[22] Wickham J D, Ritters K H, Wade T G and Vogt P 2010. A national assessment of green infrastructure and change for the conterminous United States using morphological image processing *Landsc. Urban Plan.* **94** 186–195

[23] Islam K, Jashimuddin M, Nath B and Nath T K 2018 Land use classification and change detection by using multi temporal remotely sensed imagery: The case of Chunati wildlife sanctuary, Bangladesh *Egyptian Journal of Remote Sensing and Space Science* **21** 37–47

[24] Verburg P H, Ritsema van Eck J R, de Nijs T C M, Dijst M J and Schot P 2004 Determinants of land-use change patterns in the Netherlands. *Environ. Plann. B Plann. Des.* **31** 125–150

[25] Lambin E F, Turner B L, Geist H J, Agbola S B, Angelsen A, Folke C, Bruce J W, Coomes O T, Dirzo R, Fischer G, Folke C, George P S, Homewood K, Imbernon J, Leemans R, Li X, Moran E F, Mortimore M, Ramakrishnan P S, Richards J F, Skånes H, Steffen W, Stone G D, Svedin U, Veldkamp T A, Vogel C, Xu J 2001 The causes of land-use and land-cover change: moving beyond the myths. *Global Environ. Chang.* **11** 261–269

[26] Krishnadas M, Agarwala M, Sridhara S and Eastwood E 2018 Parks protect forest cover in a tropical biodiversity hotspot, but high human population densities can limit success *Biol. Conserv.* **223** 147–155

[27] Estreguil C and Mouton C 2009 *Measuring and reporting on forest landscape pattern, fragmentation and connectivity in Europe: methods and indicators* (Italy: JRC Scientific and Technical Reports Office for Official) pp 7–8

[28] Sauders D A, Hobbs R J and Margules C R 1991 Biological consequences of ecosystem fragmentation: A review *Conserv. Biol.* **5** 18–32

[29] Fahrig L 2003 Effects of habitat fragmentation on biodiversity *Annu. Rev. Ecol. Evol. Syst.* **34** 487–515

Acknowledgments

The authors wish to acknowledge assistance and encouragement from the officers of the Department of Wildlife and National Parks (DWNP) and other Ph.D. candidates from LESTARI, UKM or financial support in terms of the scholarship received by the first author from program MyBrain 15 by the Ministry of Higher Education (MoHE).

IOP Conf. Series: Earth and Environmental Science **380** (2019) 012008 doi:10.1088/1755-1315/380/1/012008

Bacterial indicators reveal water quality status of Rangkui River, Bangka Island, Indonesia

Miratul Maghfiroh[1*], Heni Novianti[2], Lukman[1], Eddy Nurtjahya[2]

[1]Research Centre for Limnology, Indonesian Institute of Sciences (LIPI), Kompleks Cibinong Science Centre, West Java, 16911, Indonesia
[2]University of Bangka Belitung, Kepulauan Bangka Belitung, 33172, Indonesia

**Corresponding author: miratul@limnologi.lipi.go.id*

Abstract. Rangkui River flows across Pangkal Pinang City, the capital of Bangka Belitung Province. Land uses vary and can inevitably be pollution hot-spots for this river. This study aimed to investigate the current status of Rangkui River based on bacterial indicators in association with other water quality parameters. The sample collection was completed in March 2017 within 6 stations. Water quality parameters observed were temperature, TSS, turbidity, pH, DO, salinity, BOD, COD, Nitrate and Phosphate. The results of faecal coliform observation indicated that the stations closer to upstream (station 1, 2, 3) showed lower faecal coliform concentration (530-940 MPN/100 ml) as compared to those closer to downstream (station 4, 5, 6) which had the amount of ~11,000 MPN/100 ml. Phylogenetic study emphasized that isolates of interest were more similar to *Escherichia* group. The water from Rangkui River could be classified from Class I to Class IV according to Indonesia Government Regulation No.82/2001. PCA revealed Station 3 and 6 obviously separated from other stations so that more concerns should be given to improving water quality in these areas.

1. Introduction

Generally, the sources of water pollution differ and originate from point or non-point sources. Runoff of the pollutants occurs from, for instance, livestock and animal wastes, stormwater and sewer. Others may be leaked from septic system or emanated from improper treatment of industrial wastewater [1]. Polluted rivers undergo water quality changes that may include DO (Dissolved Oxygen), BOD (Biochemical Oxygen Demand), COD (Chemical Oxygen Demand), N-NH_3, TP (Total Phosphorus), pH, turbidity [2,3,4]. The value of each parameter tends to increase or decline. DO value may be lower than that of the threshold whereas other parameters such as BOD, or COD demonstrate significant increase in concentration [5,6]. Other than that, the assessment of pollution in river can be carried out by observing the content of organic compounds such as hydrocarbons [7], inorganics such as heavy metal concentration [8], physical and physiological features (colour, taste, odour, turbidity), and biological aspects (faecal coliform, enteric virus) [9,10].

Among the assessments using biological parameters, observation on faecal coliform is the most common water quality parameter to examine since the amount of this bacterial group may relatively describe biological pollution in water systems [11,12]. The occurrence of pathogenic bacteria can lead to the risk to human health through the spread of communicable disease. Pathogenic *Escherichia coli*, *Salmonella* spp, and *Cryptosporidium* spp. that are present in water exceeding government criteria

IOP Conf. Series: Earth and Environmental Science **380** (2019) 012008 doi:10.1088/1755-1315/380/1/012008

would be of community concerns [1]. The contamination of faecal coliforms in water can induce serious problems and is reported to bear negative consequences [13,14].

In Indonesia, the majority of the river has been suffering from multiple pollutions. The deterioration is caused by poor management of land and water resources which includes deforestation and improper agricultural practices [15]. In addition, more than 12% of Indonesians do not have access to sanitation facilities (i.e. toilets) [16], as it is expected that open spaces, e.g. river, are used to discharge biological wastes. This also happens in Rangkui River which has ~7,200 m in length and lies across Pangkal Pinang City, the capital of Bangka Belitung Province. The river comprises approximately 4,300 ha of catchment area. The upstream areas include Bukit Kemiri and Bukit Kelowang whereas at downstream part, Baturusa River which is adjacent to Pangkalbalam Harbour, finally discharges into the sea at the eastern part of Bangka island [17]. Rangkui River flows through three different districts which cover Kecamatan Rangkui, Kecamatan Taman Sari, and Kecamatan Pangkalbalam and along its stream, land uses vary from residential, urban commercial, recreational place, drinking water plant, traditional markets to lead mining sites. These can inevitably be pollution hot-spots for Rangkui River.

This study aimed to investigate the current status of Rangkui River based on bacterial indicators and other water quality parameters. The information also provides insights on (1) how different land use is associated with the dynamics of water quality changes; (2) potential sources of nutrient loads and bacterial emission. Additionally, the information contributes correspondingly to the sustainable management of Rangkui River by the government of Pangkal Pinang.

2. Methods

2.1 Sample collection

We collected the samples in March 2017. Water quality parameters which include temperature, turbidity, pH, salinity, DO were measured *in situ* using water quality checking devices whereas TSS (Total Suspended Solids), BOD, COD, Nitrate (N-NO_3), Phosphate (P-PO_4) were analysed in the laboratory according to standard procedures [18]. Sample collections were carried out within six sites along Rangkui River (Figure 1).

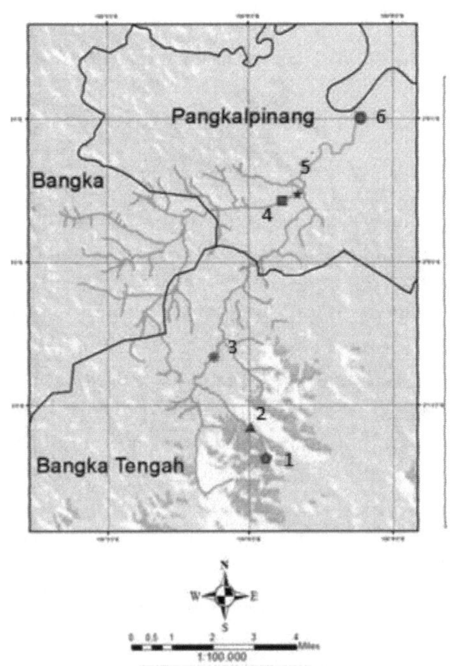

Figure 1. Location of sample collections (numbers represent the sampling stations).

Each station has specific features to represent different land use variation (table 1). Water samples were taken ~1 L each and kept in sterile bottles. The temperature was maintained at ~4^{0}C during transportation to the laboratory for further analysis.

Table 1. Site description.

	Coordinate	Attributed land use	Images
Station 1	2°13.109 S; 106 ° 06.372 E	Plantation and drinking water plant are identified in this area - upstream point in Kecamatan Pangkalan Baru	
Station 2	2°12.473 S; 106° 06.047 E	Intensive plantation site in Kecamatan Pangkalan Baru	
Station 3	2 ° 10.994 S; 106 ° 05.280 E	Metal mining spots (~50 m in length); (located in Kecamatan Pangkalan Baru)	
Station 4	2 ° 07.729 S; 106 ° 06.711 E	Urban commercial and residential of Kecamatan Rangkui	
Station 5	2 ° 07.577 S; 106 ° 07.023 E	Traditional market in Kecamatan Taman Sari	
Station 6	2 ° 05.978 S; 106 ° 08.344 E	Downstream part (Baturusa watershed) in Kecamatan Pangkalbalam	

2.2 Enumeration of faecal coliform and identification of isolate of interest

The determination of faecal coliform was done using MPN (Most Probable Number) with multiple tube method [18]. Lactose broth (LB) and Brilliant Green Lactose Bile broth (GLBB) were used respectively for presumptive/isolation and confirmatory test. Incubation was 24-48 hours at 35-37^{0}C. Triplicate was applied to provide more representative data. Isolates of interest were obtained by the observation morphological and Gram characteristics [19]. Sequential single colony isolations (SCIs) were intended to produce pure colony. EMB Agar (EMBA) was chosen for SCI routine and Nutrient Agar (NA) was the bacterial medium for cultivating purpose. DNA (deoxyribonucleic acid) from isolate of interests were extracted using DNA – GF-1 Bacterial DNA Extraction Kit (Vivantis) followed by amplification of 16S rRNA gene sequence. We used 2X Taq Pol Master Mix (Vivantis) and universal primer 9F(5'-GAGTTTGATCCTGGCTCAG) and 1492R(5'-GGCTACCTTGTTACGACTT). The setting of PCR (Polymerase Chain Reaction) as follows: 95^0C for 2 mins, 95^0C for 30 secs, 55^0C for 30 secs, 75^0C for 1 min, 75^0C for 5 mins, 20^0C for 1 min which cycled 35 time. Prior to sequencing, amplicons were purified (PCR Clean-up kit, Vivantis). Sequences were matched with those deposited in GenBank. Phylogenetic tree was constructed and analysed with MEGA 6.0 [20].

2.3 Data analysis
Data accumulated from water quality parameters and faecal coliform were then compared with the guideline published by the Government of Indonesia through Government Regulation (PP) No. 82/2001 entitled Management of Water Quality and Water Contamination Control (table 2). Thus, the status of Rangkui River is based on the criteria set by national government regulation. The abundance of faecal coliform is explained descriptively. PCA (Principal Component Analysis) was served by R programming language [21] coupled with R commander [22] and FactoMineR packages [23].

Table 2. Indonesia Government Regulation No. 82/2001 regarding water classification.

	Unit	Class I	Class II	Class III	Class IV
TSS	mg/L	50	50	400	400
pH	-	6 – 9	6 – 9	6 - 9	5 – 9
BOD	mg/L	2	3	6	12
COD	mg/L	10	25	50	100
DO	mg/L	6	4	3	0
Phosphate	mg/L	0.2	0.2	1	5
Nitrate	mg/L	10	10	20	20
Faecal Coliform	MPN/100 mL	100	1000	2000	2000

Notes:
Class I: water suitable for drinking water after further treatment
Class II: water for recreational, aquaculture, animal husbandry, farming, agriculture
Class III: water for aquaculture, animal husbandry, farming, agriculture
Class IV: water for farming, agriculture

3. Results and discussion
Rangkui River caters local people with numerous ecosystem services. This river supplies water for drinking water, recreational, animal husbandry and agriculture. Furthermore, Rangkui River is the haven for groups of fish and other aquatic organisms. We studied the status of Rangkui River to present information as the baseline data, thereby, the current condition of Rangkui River can properly be described and better comprehended.

The distribution of faecal coliform at six different sites is revealed in table 3. Faecal coliform observed in Station 1, 2, and 3 met Class II criterion of water classification Government Regulation (PP) No. 82/2001. This type of water supply is suitable for recreational, aquaculture, animal husbandry, farming, agriculture. However, faecal coliform in the rest of the stations showed more abundant numbers. Coliform contamination seems apparent for these sites (Station 4, 5, and downstream). The significant number of faecal coliforms might have been discharged to the river through waste from livestock or untreated human sewage. This finding also confirmed another result that faecal indicator organisms were more abundant at densely populated sites [24]. To eliminate more striking coliform contamination in the future, alternatively, conventional management of sewage system should be updated or replaced [25].

Table 3. Faecal coliform distribution.

	MPN/100 mL (avg.)	Attributed land use
Station 1	530	Plantation and drinking water plant are identified around this area (upstream point)
Station 2	740	Intensive plantation site
Station 3	940	Metal mining spots
Station 4	>11000	Urban commercial and residential
Station 5	>11000	Traditional market
Station 6	>11000	Downstream part (Baturusa watershed)

We also tried to observe other bacterial pathogens (presumptive) and found that *Salmonella* spp., *Vibrio* spp., *Aeromonas* sp., and *Shigella* sp. could be recovered from water sample of Station 6 (data not shown). We identified occasional colonies of bacterial pathogens (one or two bacterial groups) in other water samples apart from Station 6. This is suggested that bacterial pathogens are more predominant at downstream area.

Sequencing method allowed us to characterize our isolates of interest. Here, we selected four isolates each from designated Station 1, 3, 4 and 5. The determination is based on morphological and Gram staining. Additionally, these four isolates also demonstrated notable behaviour when challenged with antibiotic at various concentration (data not shown). UPGMA algorithm was run to generate tree construction as shown in Figure 2. Four isolates clustered within Escherichia clade. Isolate ET1 and ET4 were more similar to *Escherichia coli.*

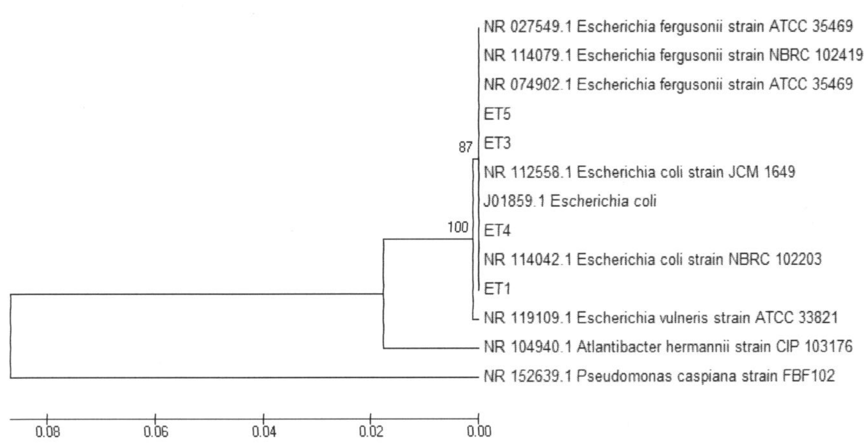

Figure 2. Phylogenetic tree based on 16S rRNA gene sequences across 1050 bp.

We also tried to describe the dynamics of water quality parameters in six stations along Rangkui River. According to table 4, The highest turbidity (also TSS) was noticed at Station 3 where the mining spot is at proximity. The lowest turbidity (and TSS) was around the upper stream area. pH was slightly below pH 7 but one pH tends to be a bit more acidic around mining area [26]. Salinity gradient got greater by the downstream part. The lowest DO was found at Station 3. Greatest concentration of BOD and COD was recognized at downstream part. This suggests that organic compounds were predominantly accumulating at downstream. Nitrates may be associated with agriculture practices. Nitrates from fertilizer can be carried through rain and contaminate the water [27]. Nitrates from agriculture contribute to the elevated level of nitrate at upstream site. Phosphates at high level can trigger the bloom of algae. At Station 6, the amount of phosphates should be of concern since it was observed abundantly.

Table 4. Observations of water quality parameters.

		Station					
	Unit	1	2	3	4	5	6
Temp.	^{0}C	25,17±0,29	27,20±0,29	31,17±0,29	28,00±1,23	29,50±0,25	30,00±1,48
Turbidity	NTU	1,55±0,16	1,19±0,17	408,33±10,5	33,44±21,27	86,00±4,00	88,00±5,20
TSS	mg/L	0,06±0,01^a	0,04±0,02^a	15,93±1,84^a	1,31±0,47^a	3,44±0,95^a	3,52±0,24^a
pH	-	6,54±0,03^a	6,63±0,06^a	5,51±0,17^d	6,71±1,03^a	6,45±0,49^a	6,86±0,02^a
Salinity	Ppt	2,00±1,09	1,50±0,43	1,50±0,25	2,00±0,75	2,50±0,43	12,00±1,09
DO	mg/L	5,39±0,83^b	4,88±0,4^b	2,91±0,12^d	3,63±0,59^c	3,06±0,11^c	3,01±1,34^c
BOD	mg/L	1,20±0,08^a	1,47±0,23^b	2,14±0,47^a	1,05±0,39^a	3,41±0,91^b	3,80±0,60^b
COD	mg/L	2,16±0,05^a	7,21±2,46^a	9,41±3,62^a	4,94±0,31^a	8,38±0,96^a	12,90±1,49^b
Nitrate	mg/L	1,90±0,36^a	1,89±0,12^a	1,30±0,15^a	1,70±0,74^a	1,90±0,79^a	1,70±1,04^a
Phosphate	mg/L	0,07±0,05^a	0,05±0,01^a	0,17±0,02^a	0,08±0,01^a	0,09±0,01^a	0,89±0,03^c

Notes: superscripted letters denote that the value meets criterion of water classification (PP No.82/2001) from: (a) Class I (b) Class II (c) Class III (d) Class IV

Water classification according to PP No. 82/2001 comprises of four Classes which represent the suitable purposes of water to be used. Water quality parameters at Station 1 and 2 demonstrated that the waters are classified to Class I or Class II. At Station 4, 5, and 6, the waters might be classified to Class III if we consider the low level of DO and the high content of phosphates at Station 6. The water at Station 3 displayed poor quality of DO and pH, thus, the water seems to be suitable for farming and agriculture (Class IV). To mitigate worst water quality in the future, improvement should be addressed. The improvement of water quality can be done by water planning and better management via public participation [28].

PCA, as defined in Figure 3, revealed distinct clustering among stations. Station 3 and 6 obviously separated from other stations. At Station 6 (downstream), it is apparent that associated water qualities were the abundance of faecal coliform, the concentration of phosphates, BOD, COD, and salinity. In Station 3, it was clearly associated with great concentration of TSS and turbidity. Since these two stations relatively had poor water quality performance, more concerns should be given to improving these areas.

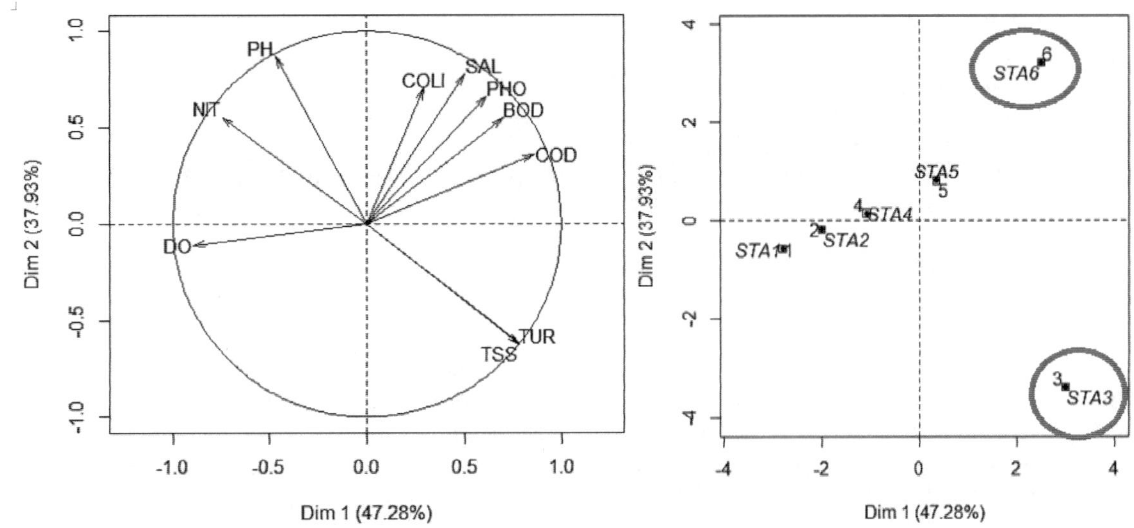

Figure 3. PCA built from water quality variables and stations as factors.

4. Conclusion

This paper identified current water quality status of Rangkui River with special reference to faecal coliform as bacterial indicators in association with other physical and chemical parameters of the river. Faecal coliform, BOD, COD, turbidity, TSS, Phosphates were a set of limiting parameters that distinguished the areas with poor water quality performance. We recommend the improvement of water quality by which water planning and better management can be implemented via public participation. Long term monitoring of water quality is indispensable to ensure the progress of its improvement.

5. References

[1] Culbertson CW, Huntington TG, Stoeckel DM, Caldwell JM, O'Donnell C 2014 *Water quality and sources of faecal coliform bacteria in the Meduxnekeag River, Houlton, Maine* (USA: U.S. Geological Survey Scientific Investigations Report) 2013–5144 p 31

[2] Effendi H, Romanto, Wardianto Y 2015 1st *Int. Symp. on LAPAN-IPB Satellite for Food Security and Environmental Monitoring* (Indonesia: Procedia Environmental Sciences Elsevier) 24 pp 228 - 237

[3] Effendi H 2016 2nd *Int. Symp. on LAPAN-IPB Satellite for Food Security and Environmental Monitoring* (Indonesia: Procedia Environmental Sciences Elsevier) 33 pp 562 - 567

[4] Hoseinzadeh E, Khorsandi H, Wei C and Alipour M 2015 *Desalination and Water Treatment* **54** 2994

[5] Bhuyan MS, Bakar MA, Sharif ASM, Hasan M and Islam MS 2018 *Pollution* **4** 481

[6] Kido M, Yustiawati, Syawal MS, Sulastri, Hosokawa T, Tanaka S, Saito T, Iwakuma T and Kurasaki M 2009 *Environ Monit Assess* **156** 317

[7] Villa RS, Doque WO, Paez M and Schumacher M 2016 *Science of the Total Env* **540** 455

[8] Islam MS, Ahmed MK, Raknuzzaman M, Mamun MH and Islam MK 2015 *Ecological Indicators* **48** 282

[9] Al-Badaii F and Shuhaimi-Othman M 2015 *Water Qual Expo Health* **7** 319

[10] Marie V and Lin J 2017 *J Water Health* **15** 966

[11] Mallin MA, Mclver MR, Robuck AR and Dickens AK 2015 *Water Air Soil Pollut* **2015** 226

[12] Craig SG 2016 *Bacterial Indicators of Fecal Pollution: Exploring Relationships between Fecal Coliforms and Enterococcus Groups in Central and South Florida Surface Waters* (Master's thesis: Nova Southeastern University Retreived from NSU Works) 445

[13] Lekshmi M, Das O, Kumar S and Nayak BB 2018 *Veterinary World* **11** 1285

[14] Kapaya F *et al* 2018 *The Pan African Medical Journal* **29**
[15] Partowijoto A 2006 *Monitoring and studies on water quality in some rivers and related water bodies in Indonesia* (Indonesia: 2nd NARBO General Meeting Jatiluhur)
[16] Indonesia Ministry of Health 2013 *Laporan Nasional Riset Kesehatan Dasar (Riskesdas)* (Jakarta: Ministry of Health)
[17] Balai Pengendalian Dampak Lingkungan Daerah (Bapedalda) Pangkal Pinang 2004 *Pemantauan Kualitas Air Sungai Rangkui* (Pangkal Pinang: Bapedalda Provinsi Kepulauan Bangka Belitung)
[18] American Public Health Association 2005 *Standard methods for the examination of water and wastewater* 21st Ed (New York: APHA)
[19] Benson, HJB and Alfred E 2010 *Benson's microbiological applications: laboratory manual in general microbiology* (New York : McGraw-Hill Higher Education)
[20] Tamura K, Dudley J, Nei M and Kumar S 2007 *Molecular Biology and Evolution* **24** 1596
[21] R Core Team 2016 *R: A language and environment for statistical computing* (Austria: R Foundation for Statistical Computing)
[22] Fox J 2005 *Journal of Statistical Software* **14** 1
[23] Le S, Josse J and Husson F 2008 *Journal of Statistical Software* **25** 1
[24] Johannessen GS, Wennberg AC, Nesheim and Tryland I 2015 *Int. J. Environ. Res. Public Health* **12** 6979
[25] Rogers K 2016 *Reduction of faecal coliform bacteria through the elimination of sewage discharges in West Virginia Streams* (Thesis and Dissertation-Public Health: University of Kentucky) p 79
[26] Tordoff GM, Baker AJM and Willis AJ 2000 *Chemospheren* **41** 219-228
[27] Sebilo M, Mayer B, Nicolardot B, Pinay G, Mariotti A 2013 *Proceedings of the National Academy of Sciences* 110 18185
[28] Bohnet IC 2015 *Society & Natural Resources* **28** 180

Bifurcation analysis of the prey-predator models incorporating herd behaviours

Z I A Manaf[1,2*] and M H Mohd[2*]

[1]Faculty of Computer and Mathematical Science, Universiti Teknologi MARA, 18500 Machang, Kelantan Darul Naim, MALAYSIA.
[2]School of Mathematical Science, Universiti Sains Malaysia, 11800, USM Penang, Malaysia.

**Corresponding authors: zati310@gmail.com and mohdhafizmohd@usm.my*

Abstract: In ecology, herd behaviour corresponds to the situation how a group of animals of the same species interacts in the time of danger. A herd mechanism offers more protection for preys against predators. But this mechanism can also provide defence from predators. To address this problem, we consider two ecological models of prey-predator interactions with the presence of different herd behaviours. While the first system represents a simple prey-predator model incorporating herd behaviour in prey populations only, the second one considers the herd behaviour in both prey and predator populations. We examine how different strategies of herd behaviours can determine the outcomes of species interactions. To do this, we employ stability analysis and numerical simulations to illustrate different dynamical behaviours of the two models, e.g. coexistence and extinction steady states and also limit cycles. By using bifurcation analysis, we observe the occurrences of Hopf bifurcation, transcritical bifurcation, limit point and limit point of cycles. These bifurcations can shape the overall dynamics of these two ecological systems.

1. Introduction

Herd behaviour corresponds to the situation how a group of animals of the same species interact in the time of danger. The preys will employ various defensive mechanisms to save themselves from predator [1]. For example, when animals are in herd, they will move together at the same time; one or two leaders start, then momentum builds as more and more will join until a large group are all heading in the same direction. Another example of herd behaviour mechanism employed by preys: when they are approached by predators, they will move together as close as possible to the centre of the fleeing group with a hope of reducing danger to themselves [2]. The larger the prey population is, the smaller the success of hunting by predators [3]. However, predators can also employ a herd behaviour mechanism to capture these individual preys. For example, wolves will usually live and hunt in family groups so that they can pursue larger prey. Besides that, predators live in group because they can protect their little ones from another large predator. Therefore, the main purpose of this paper is to get better understanding about the herd behaviour mechanisms in both prey-predator populations. To achieve this objective, we examine two ecological models of prey-predator interactions with herd behaviours using bifurcation analysis and techniques from dynamical systems.

In general, ecological models are developed to understand how population abundance of predators affects each individual prey. Various authors have studied the herd behaviour mechanisms in prey-predator interactions [1,4,5,6]. Braza [6] proposed the prey-predator models in which the prey exhibits herd behaviour and the system leads to square root of prey density. The following system given by

$$\frac{dX}{dT} = rX\left(1-\frac{X}{K}\right) - \frac{\alpha\sqrt{X}Y}{1+T_h\alpha\sqrt{X}}$$

$$\frac{dY}{dT} = -\delta Y + \frac{\beta\alpha\sqrt{X}Y}{1+T_h\alpha\sqrt{X}} \tag{1}$$

where X and Y denote the density of the prey and predator, respectively. It is assumed that all parameters are constant and positive where the parameter r is the growth rate of prey, K is the carrying capacity of the prey, α represents the rate which predator destroys prey or attack rate, δ represents the death rate of predator, T_h is the average handling time and β stand for biomass conversion rate. For ecological model (1), Braza assumed that the average handling time is zero and this leads to a simple Lotka-Volterra system.

In this paper, our intention is to establish the different effects of herd behaviours in prey-predator populations as the magnitude of some ecologically-relevant parameters change. In particular, we seek to answer the following questions: when does the herd behaviours matter in determining the dynamics of species interactions change and when does this phenomenon depends on ecological process e.g. conversion rate of prey?. This paper only focuses on the parameter of the conversion rate of prey; where predators attacking and consuming preys and hence preys converts into new-born of predators. We chose to vary that parameter because we want to investigate what happen to the prey and predator population when the herd behaviour mechanisms take place in the two different ecological models; while the first model employs the herd behaviour in prey populations only and the other model employs the herd behaviour in both prey and predator populations. Since preys live in herd (as a protection against predators), then predators might hesitate to attack preys. But then, if the predators also live in herd, it might be more effective at pulling down a herd of prey. Therefore, these situations are very interested to study. From ecological hypothesis, if the conversion rate of prey increases, this can lead to the extinction of prey and survival of predators. These situations or hypothesis will be explained and discussed further in Section 4 and 5.

This paper is organized as follows. Section 2 discusses a simple prey-predator model incorporating herd behaviour in prey populations only. Section 3 highlights the prey-predator model with herd behaviour in both prey and predator populations. Both models are extended version, the first one proposed by Bera et al. [5], another one proposed by Maiti et al. [4]. Existence of equilibrium and stability is discussed in Section 2 and 3 simultaneously. Numerical simulations and bifurcation analysis are demonstrated in Section 4. Section 5 contains the general discussion and the conclusion of the paper.

2. A Model with Herd Behaviour in Preys

To model herd behaviours in prey populations, some studies [5,6] proposed a deterministic model (1) with prey-predator interactions. This model can be non-dimensionalized with the following scaling:

$$x = \frac{X}{K}, \quad y = \frac{Y}{K}, \quad t = rT$$

System (1) then takes the form:

$$\frac{dx}{dt} = x(1-x) - \frac{b\sqrt{x}y}{1+a\sqrt{x}} \quad , x(0) > 0 \quad ,$$

$$\frac{dy}{dt} = -dy + \frac{c\sqrt{x}y}{1+a\sqrt{x}} \quad , y(0) > 0 \quad , \tag{2}$$

where $a = T_h \alpha \sqrt{K}$, $b = \frac{\alpha\sqrt{K}}{r}$, $c = \frac{\beta\alpha\sqrt{K}}{r}$, $d = \frac{\delta}{r}$.

Parameters a represent the time spent of predator in handling one prey, b is the attack rate at which predator captures their prey, c is the conversion rate each individual prey into new-borns of predators and d stand for death rate of predator. Positivity and boundedness of the system are discussed in [5]. The system always has the trivial equilibrium $E_0 = (0,0)$ where both prey and predator populations can go extinct, a boundary equilibrium $E_1 = (1,0)$ where only prey survives and the interior equilibrium $E^* = (x^*, y^*)$ where both prey and predator populations coexist with $x^* = \frac{d^2}{(c-ad)^2}$ and $y^* = \frac{cd\{(c-ad)^2 - d^2\}}{b(c-ad)^4}$.

The stability at $E_0 = (0,0)$ cannot be evaluated since the system (2) is not linearizable at $(0,0)$; this is due to the square root term and the Jacobian matrix becomes indeterminate, to analyse the stability of E_0, one can proceed to use the method proposed in [5]. At $E_1 = (1,0)$, the system (2) is stable if $c < d\,(1+a)$. Next, we consider the stability of the interior equilibrium $E^*(x^*, y^*)$ by the given Jacobian matrix

$$J(E^*) = \begin{bmatrix} a_{11} & a_{12} \\ a_{21} & 0 \end{bmatrix},$$

where

$$a_{11} = \frac{(c+ad)(c-ad)^2 - (ad+3c)d^2}{2c(c-ad)^2}$$

$$a_{12} = -\frac{bd}{c}$$

$$a_{21} = \frac{(c-ad)^2 - d^2}{2b(c-ad)}.$$

If $(ad+3c)d^2 - (c+ad)(c-ad)^2 > 0$, then $E^*(x^*, y^*)$ is locally asymptotically stable. To prove this statement, we have the following characteristic equation of $J(E^*)$

$$\lambda^2 + A\lambda + B = 0, \tag{3}$$

where

$$A = -tr\, J(E^*) = -a_{11} - a_{22} = \frac{(c+ad)(c-ad)^2 - (ad+3c)d^2}{2c(c-ad)^2}$$

$$B = -\det J(E^*) = a_{11}a_{22} - a_{12}a_{21} = \frac{d((c-ad)^2 - d^2)}{2bc(c-ad)}.$$

For existence of $E^*(x^*, y^*)$, we have $d < \frac{c}{1+a}$ that confirms $B > 0$ and $A > 0$ also. Thus, the root of (3) is given by

$$\lambda_{1,2} = \frac{-A \pm \sqrt{A^2 - 4AB}}{2},$$

with negative real part and this guaranteed the stability of $E^*(x^*, y^*)$.

3. A Model with Herd Behaviour in Both Preys and Predators

The prey-predator system is originally formulated by Maiti et al. [4]. The authors assumed that both prey and predator population live in herds and square root term are employed in both prey and predator density. The following system given by:

$$\frac{dx}{dt} = x(1-x) - \frac{b\sqrt{x}\sqrt{y}}{1+a\sqrt{x}}$$

$$\frac{dy}{dt} = -dy + \frac{c\sqrt{x}\sqrt{y}}{1+a\sqrt{x}} \tag{4}$$

where $a = T_h\alpha\sqrt{K}$, $b = \frac{\alpha}{r}$, $c = \frac{\beta\alpha}{r}$, $d = \frac{\delta}{r}$.

The stability at $E_0 = (0,0)$ and $E_1 = (1,0)$ cannot be evaluated since the system (4) are not linearizable at $(0,0)$ and $(1,0)$; this is due to the square root term and the Jacobian matrix becomes indeterminate, to analyse the stability of E_0 and E_1, one can proceed to use the method proposed as in [4]. The stability of the interior or coexistence equilibrium $E^* = (x^*, y^*)$ are determined by the given Jacobian matrix

$$J(E^*) = \begin{bmatrix} a_{11} & a_{12} \\ a_{21} & a_{22} \end{bmatrix},$$

where

$$
\begin{aligned}
a_{11} &= 1 - 2x^* - \frac{b\sqrt{y^*}}{2\sqrt{x^*}(1 + a\sqrt{x^*})^{2'}} \\
a_{12} &= -\frac{b\sqrt{x^*}}{2\sqrt{y^*}(1 + a\sqrt{x^*})'} \\
a_{21} &= \frac{c\sqrt{y^*}}{2\sqrt{x^*}(1 + a\sqrt{x^*})^{2'}} \\
a_{22} &= -\frac{d}{2}.
\end{aligned}
$$

The characteristics equation of $J(E^*)$ is given by

$$\lambda^2 + C\lambda + D = 0,$$

where

$$
\begin{aligned}
C &= -tr\ J(E^*) = -a_{11} - a_{22} \\
D &= -\det J(E^*) = a_{11}a_{22} - a_{12}a_{21}.
\end{aligned}
$$

The stability of interior equilibrium is discussed further in [4].

4. Numerical Bifurcation Analysis

A bifurcation analysis has been carried out to demonstrate different dynamical behaviours of the models in Section 2 and 3 as conversion rate c is varied. In particular, we perform some simulation using MatCont (MATLAB continuation package) for the models (2) and (4) with parameter values as in table 1.

Table 1. Parameter used for both ecological models (2) and (4).

Parameter	Definition	Value	Source
a	Handling time	0.4	[2]
b	Attack rate	0.15	[2]
c	Conversion rate	0.85	[4]
d	Death rate	0.5	[4]

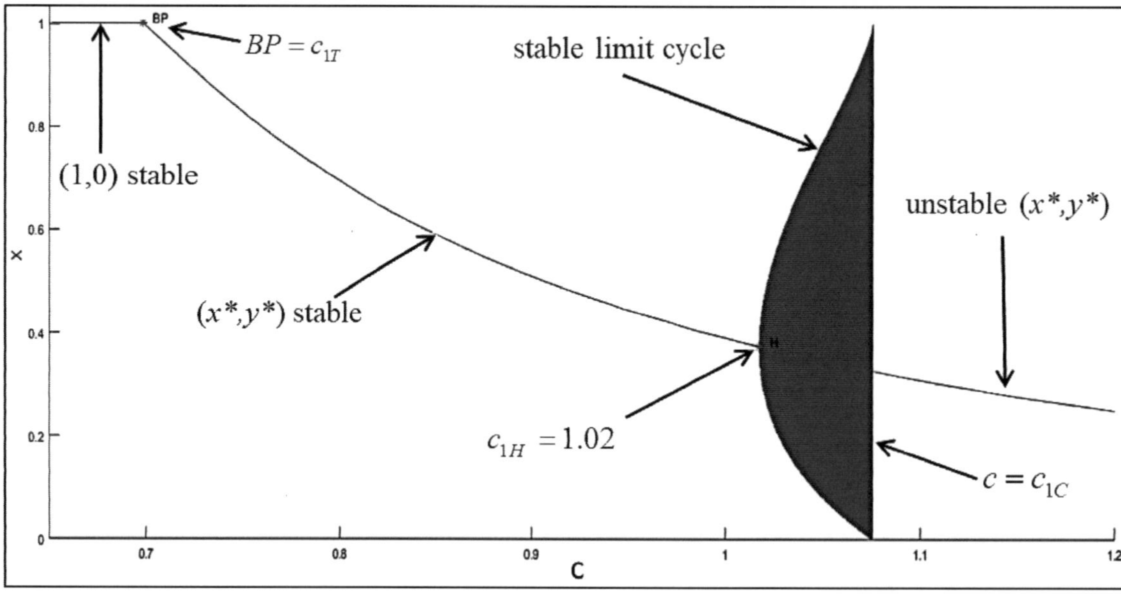

Figure 1. Bifurcation diagram of system (2) as c (conversion rate) varies versus prey populations for $0 < c < 1.2$.

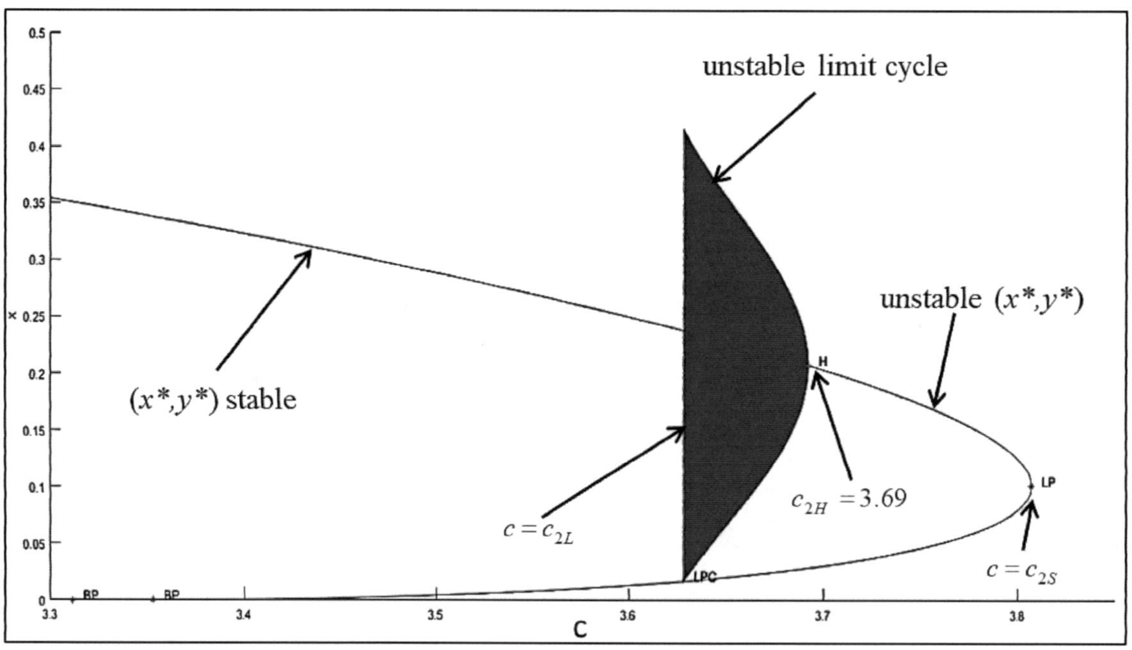

Figure 2. Bifurcation diagram of system (4) as c (conversion rate) varies versus prey populations for $3.3 < c < 3.8$.

Using bifurcation analysis (figure 1), we observed different types of behaviour as bifurcation parameter c changes. For $c < c_{1T}$, single-species steady state (1,0) is stable: preys are free from predation and hence achieved the carrying capacity of the population since the predators go extinct in this case. At $c = c_{1T}$, we observed the occurrences of transcritical bifurcation (BP). There is an exchange of stability between single-species steady state (1,0) with two-species steady state (x^*, y^*). When

$c_{1T} < c < c_{1H}$, two-species steady state is stable. At $c = c_{1H}$, supercritical Hopf bifurcation (HB) occurs. For $c_{1H} < c < c_{1C}$, the two-species steady state becomes unstable with corresponding creation of stable limit cycles. As parameter c increases, the stable limit cycles continue to grow in amplitude until this cycle collides with a saddle point at $c = c_{1C}$; this phenomenon is called homoclinic bifurcation and this occurs when $c = c_{1C}$. As $c > c_{1C}$, the limit cycles disappears completely. Thus, this situation causes the preys to go extinct, then followed by the exclusion of predators.

Figure 2 shows the bifurcation analysis for ecological model (4) when parameter c changes. When $c_{2H} < c < c_{2S}$, two-species steady state is unstable. At $c = c_{2S}$, the limit point (LP) occurs whenever at $c = c_{2H}$,subcritical Hopf bifurcation occurs. For $c_{2H} < c < c_{2L}$ the two-species steady state becomes stable with corresponding creation of unstable limit cycles. As parameter c decreases, the unstable limit cycles continue to grow in amplitude until this cycle meets the limit point of cycle (LPC) when $c = c_{2L}$. As $c < c_{2L}$, the limit cycles disappears completely and both prey and predator populations coexist since the two-species steady state is stable.

Next, three different values (see table 2 and 3) of bifurcation parameter c are chosen to describe the dynamics of the prey-predator systems with herd behaviour as bifurcation parameter changes. The phase portraits and time series graphs for each values of c are plotted (see figure 3-8).

Table 2. Stability and bifurcation analysis results of model (2).

Bifurcation parameter	Steady states $E^* = (x^*, y^*)$	Eigenvalues	Characteristics	Figure
$c = 0.95$	$(0.4444, 3.1276)$	$\lambda_1 = -0.0541 + 0.3267i$ $\lambda_2 = -0.0541 - 0.3267i$	Stable steady state	3(a) 3(b)
$c = 1.069$	$(0.3311, 3.1565)$	$\lambda_1 = 0.0330 + 0.3672i$ $\lambda_2 = 0.0330 - 0.3672i$	Stable limit cycle (unstable steady state)	4(a) 4(b)
$c = 1.15$	$(0.2770, 3.0709)$	$\lambda_1 = 0.0737 + 0.3793i$ $\lambda_2 = 0.0737 - 0.3793i$	Unstable steady state (stable steady state at $E_0(0,0)$ only)	5(a) 5(b)

Table 3. Stability and bifurcation analysis results of model (4).

Bifurcation parameter	Steady states $E^* = (x^*, y^*)$	Eigenvalues	Characteristics	Figure
$c = 3.55$	$(0.2700, 9.3292)$	$\lambda_1 = -0.0461 + 0.1843i$ $\lambda_2 = -0.0541 - 0.1843i$	Stable steady state	6(a) 6(b)
$c = 3.6285$	$(0.2378, 8.7689)$	$\lambda_1 = -0.0222 + 0.1669i$ $\lambda_2 = -0.0222 - 0.1669i$	Unstable limit cycle (stable steady state)	7(a) 7(b)
$c = 3.75$	$(0.173, 7.1590)$	$\lambda_1 = 0.0246 + 0.1148i$ $\lambda_2 = 0.0246 - 0.1148i$	Unstable steady state (stable steady state at $E_0(0,0)$ only)	8(a) 8(b)

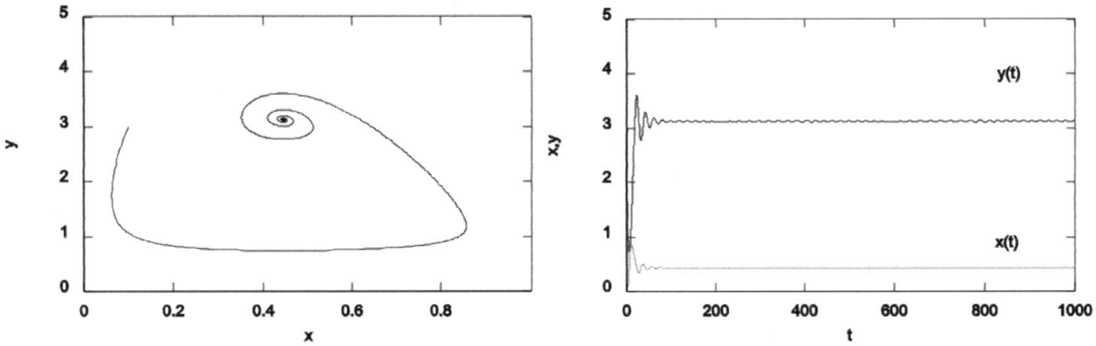

(a) (b)

Figure 3. Phase portrait (a) and time series graph (b) for system (2) with $c = 0.95$ and initial conditions $x(0) = 0.1$ and $y(0) = 3$. The equilibrium point E^* is a stable focus (a).The prey $x(t)$ and predator $y(t)$ populations converge to an equilibrium value (b).

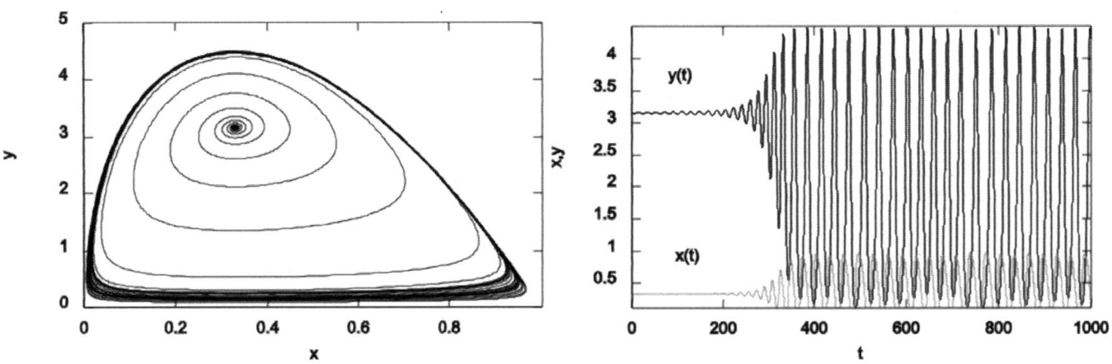

(a) (b)

Figure 4. Phase portrait (a) and time series graph (b) for system (2) with $c = 1.069$ and initial conditions $x(0) = 0.33$ and $y(0) = 3.16$. The equilibrium point E^* is an unstable focus and there is a stable periodic solution arises as time (t) increases.

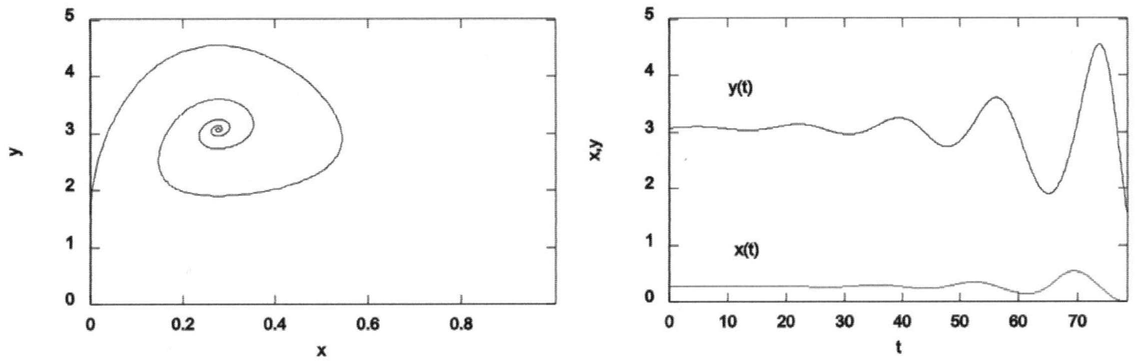

(a) (b)

Figure 5. Phase portrait (a) and time series graph (b) for system (2) with $c = 1.15$ and initial conditions $x(0) = 0.28$ and $y(0) = 3.07$. The equilibrium point E^* is an unstable focus (a). The prey $x(t)$ and predator $y(t)$ populations diverge from an equilibrium value (b).

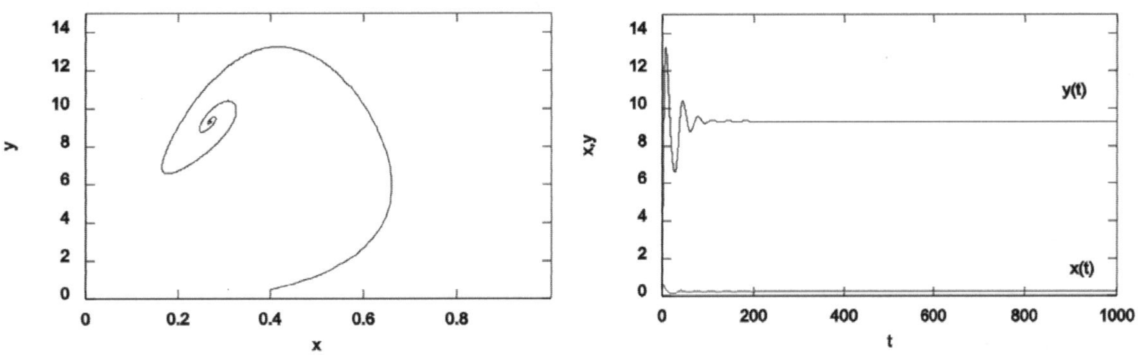

(a) (b)

Figure 6. Phase portrait (a) and time series graph (b) for system (4) with $c = 3.55$ and initial conditions $x(0) = 0.4$ and $y(0) = 0.5$. The equilibrium point E^* is a stable focus (a). The prey $x(t)$ and predator $y(t)$ populations converge to an equilibrium value (b).

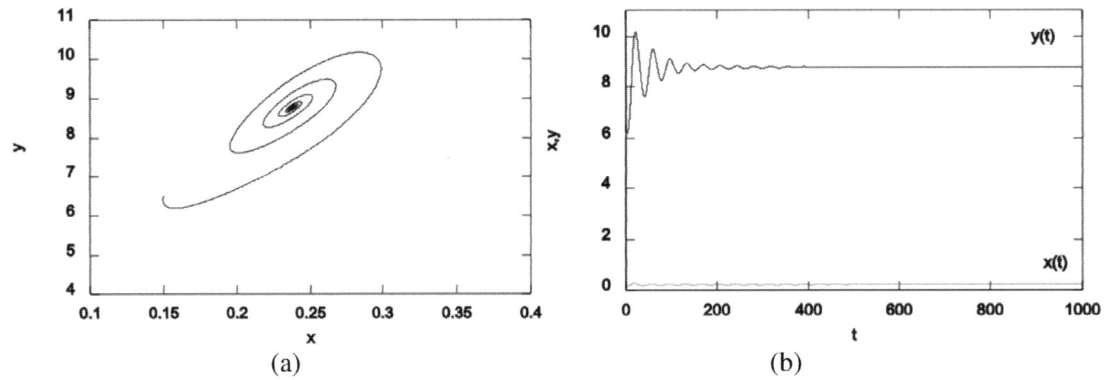

Figure 7. Phase portrait (a) and time series graph (b) for system (4) with $c = 3.6285$ and initial conditions $x(0) = 0.15$ and $y(0) = 6.5$. The equilibrium point E^* is a stable focus (a).

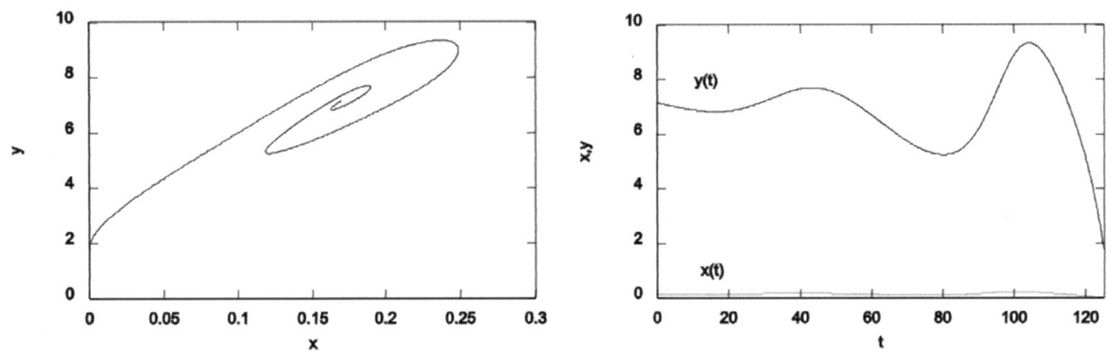

Figure 8. Phase portrait (a) and time series graph (b) for system (4) with $c = 3.75$ and initial conditions $x(0) = 0.17$ and $y(0) = 7.15$. The equilibrium point E^* is an unstable focus (a). The prey $x(t)$ and predator $y(t)$ populations diverge from an equilibrium value (b).

5. Discussion and ecological implications

In this paper, the dynamics of the two ecological models of prey-predator is studied to understand the joint effects of herd behaviour mechanism and prey conversion rate in determining the population dynamics of species. Correspondingly, the ecological model (2) employs the herd behaviour in prey populations and the other model (4) employs the herd behaviour in both prey and predator populations. The stability and bifurcation results of each bifurcation parameter c are carried out by using mathematical continuation software i.e. in MATCONT and XPPAUT.

By using bifurcation analysis (see table 2, figure 1), we observed the occurrences of Supercritical Hopf and Homoclinic bifurcation on the model (2). Homoclinic bifurcation occurs when a periodic orbit collides with a saddle point [7]. As bifurcation parameter c increases, the limit cycle grows until it exactly intersects the saddle point, yield an orbit of infinite duration but if parameter increases further, the limit cycle disappears completely (see figure 1). We also observed the occurrences of Subcritical Hopf bifurcation on the model (4). It occurs when the stability of the equilibrium changes via a pair of purely imaginary eigenvalues and will result in unstable limit cycles [8] (see figure 2).

Ecologically, in comparison to the herd behaviour mechanisms of the two ecological models, the value of parameter c are discussed and summarized into three categories i.e. the low, moderate and high value of c. The low value of c shows an extinction of predator in model (2); since the preys live in herd, therefore the predators might hesitate to attack preys. However, when the predators also live in herd, it is very effective for them to pull down a herd of prey, and this will result the coexistence of two-species in model (4). For moderate value of c , both models (2) and (4) shows two-species coexistence; the prey and predator populations never goes to an extinction (see figure 3, figure 6). But then, in model (2), oscillatory solutions emerged due to Hopf bifurcation, where the prey and predators start to oscillate as the time changes (see figure 4). At this point, the preys oscillate to a very low population density will be excluded entirely and consequently, these situations can cause destabilized of coexistence of the two-species. For this reason, as the high value of c , there is an extinction of two-species in both models (2) and (4); the prey population first goes to extinction and then the predator population follows suit (see figure 5, figure 8).

In short, this paper illustrates the significant roles of herd behaviour mechanisms in the dynamics of prey-predator systems. By introducing herd behaviour mechanisms in predator populations, it helps the predator to survive themselves in nature and subsequently, the two-species coexist and this can increase the species diversity. For future research, it is interesting to explore more about the herd behaviour mechanisms, such as comparing the two ecological models in terms of two-parameter bifurcations analysis.

6. References

[1] Matia S N and Alam S 2013 Prey-predator Dynamics under Herd Behavior of Prey *Universal Journal of Applied Mathematics* **1**(4) 251–257

[2] Hamilton W D 1971 Geometry for the selfish herd *Journal of Theoretical Biology* **31**(2)295-311

[3] Tener J S 1965 Muskoxen in Canada a biological and taxonomic review *Canadian Wildlife Service Monograph 2* (Ottawa: Queen's Printer)

[4] Maiti A, Sen P and Samanta G P 2016 Deterministic and stochastic analysis of a prey – predator model with herd behaviour in both *System Science and Control Engineering* **4**(1) 259-269

[5] Bera S P, Maiti A and Samanta G P 2015 Modelling herd behavior of prey : analysis of a prey-predator model *World Journal of Modelling and Simulation* **11**(1) 3-14

[6] Braza P A 2012 Prey – predator dynamics with square root functional responses *Nonlinear Analysis: Real World Applications* **13**(4) 1837–1843

[7] Strogatz S H 1994 *Nonlinear Dynamics and Chaos* (Canada: Perseus Books Publishing) 262

[8] Andronov A A, Leontovich E A, Gordon I I and Maier A G 1971 Theory of Bifurcations of Dynamical Systems on a Plane (Jerusalem: Israel Program for Scientific Translations)

Acknowledgements

Thanks to the School of Mathematical Sciences and the Universiti Sains Malaysia for the support. Mohd Hafiz Mohd is supported by the Universiti Sains Malaysia (USM) Fundamental Research Grant Scheme (FRGS) No. 203/PMATHS/6711645.

Larvae Density and Environmental Condition as Risk Factors to Dengue Incidence in Semarang City, Indonesia

M Martini[1,2,*], J Annisa[1], LD Saraswati[1], R Hestiningsih[1], N Kusariana[1] and S Yuliawati[1]

[1] Epidemiology and Tropical Disease Department, Public Health Faculty Of Diponegoro University, Jl. Prof Sudarto Tembalang Semarang, Indonesia, 50275
[2] Magister Epidemiology, Pasca sarjana School, Imam Bardjo street, Semarang, Indonesia 50241

**Corresponding author: tinihen65@yahoo.co.id*

Abstract. Dengue Hemorrhagic Fever (DHF) is a disease caused by a virus transmitted by a mosquito vector of *Aedes*. As Semarang city of Indonesia has dense population and is located in a low altitude area, Semarang is identified to be a high endemic area of DHF. The objective of this study was to describe the risk factors of DHF, specifically larva density and environmental condition, using spatial analysis. The descriptively observational research with cross sectional approach was used and samples were taken using proportional stratified random sampling technique in Tembalang sub-district on 100 households. Data obtained through questionnaires, observations, and measurements were analyzed by spatial method. The result showed that the description of the risk factor was as follows: larva density (CI) was still above the national standard of Indonesia by 5%. The behavior of respondent of Semarang city belonged to the category of poor, and the majority of respondents were not routinely doing elimination of mosquito breeding sites.

1. Introduction

Dengue Hemorrhagic Fever (DHF) and Dengue Shock Syndrome (DSS) is a disease caused by a virus transmitted through mosquito vectors, and spreads quickly [1]. Three types of diseases caused by dengue virus are Dengue Fever (DD), DHF, and DSS. Dengue virus belongs to group B Arthropod Virus (Arbovirosis) that has 4 serotypes: Den-1, Den-2, Den-3, and Den-4 [2,3]. Every year, it is estimated that there are about 390 million dengue infections worldwide, from which three quarters of DHF sufferers are in the Asia-Pacific region [4].

In Central Java Province, Indonesia, DHF is still a health problem. The city of Semarang, of Central Java Province, is an endemic area and has the highest cases compared to the cases in other districts/cities. The incidence of DHF in Central Java Province in 2016 was 43.4 per 100,000 population, and the Case Fatality Rate was 1.46% [5,6]. Meanwhile, the incidence of DHF in Semarang City was 98.61 per 100,000 population in 2015, and 25.22 per 100,000 inhabitants in 2016 [6]. In the Semarang City, the highest incidence of DHF was identified in Tembalang sub-district as many as 194.04 per 100,000 population in 2015, and 207.57 per 100,000 population in 2016 [7]. The data report in Dengue Program in Indonesia, DHF and DSS make one data as a DHF cases.

Epidemiological studies of DHF reveal that the factors associated with DHF are host factors including age, immunity status, and nutritional status; agent factor that is dengue virus, which is transmitted through *Aedes aegypti* and *Aedes albopictus* mosquito vectors; while, environmental factors are geographic conditions (height from the sea level, rainfall, wind speed, humidity, season/climate) and demography [8,9]. Other factors proven to play a role in increasing the spread of dengue cases are vector density, continually increasing population, uncontrollable urbanization, an increase in transportation facilities, unhealthy community behavior towards the environment, as well as climate change [4,10].

This study examined the factors of *Aedes* sp vector and environment related to the incidence of DHF in Tembalang sub district, Semarang, Indonesia. This study was conducted using spatial analysis; an analysis and description based on geographical disease data [2]. Utilization of spatial analysis of the incidence of DHF is expected to provide benefits in describing the pattern of the spread of DHF in a region.

2. Materials and methods

This descriptive study with a cross sectional approach was conducted in one of the sub-districts in Semarang City, namely Tembalang sub district, which is considered the DHF endemic district as the highest incidence of DHF has been identified in this region.

The research sample was 100 households registered as permanent residents' domicile in 12 villages under the administration of Tembalang sub district. To describe the distribution of DHF and spatial environmental factors, village unit was used for the 12 villages in Tembalang sub district. The respondents were the head of the family or housewives. The variables measured were the DHF incidence, larvae density, environmental parameters such as temperature and humidity, population density, and population behavior in eradicating mosquito breeding sites (EMB). Data were analyzed descriptively and displayed in the form of maps and tables.

3. Results and discussion

3.1. Larvae Density and Environmental Factors in Tembalang Sub District, Semarang

The results of the entomological survey showed that the larva density was quite high in Tembalang sub district as measured by Container Index (CI) with the average larvae density or the CI was 9.83% per village; while, the national standard was less than 5%. The CI is calculated by the percentage of the positive number of containers divided by the number of containers examined. From the perspective of standard program of vector control, the larvae density in the villages of Tembalang sub district was mostly dense (8 villages from 12 villages were identified to be having high larvae density). High density is categorized when the CI in a region is 5% or more (standard of national vector control program) [6,7].

The types of containers mostly observed were bathtubs, some were made from ceramic, and ceramics are easier to clean and not easily to be mossy (48%); while, the least containers to be observed were made of cement (20%). The larvae found in the containers were further identified their species in the entomology laboratory of Diponegoro University, and the results showed that all (100%) were identified as *Ae. aegypti*.

Physical environmental parameters observed were water pH, temperature, and humidity. Ninety percent of the water used by respondents had the average water pH of 6.812. Meanwhile, the environmental air temperature around the respondents' house was in usual interval (25°-35°C), and the average humidity was 53.95% (indoor) and 53.35% (outdoor). The average altitude of the respondents' residence was 92.16 meter above the sea level.

The social environment associated with population density showed that the average population density of Tembalang sub district was 7,392 people/km^2, with the most densely populated was Jangli Village with 10,876 people/km^2. Respondents' MBE behaviour was still low (82%) (table 1). Many respondents left their water shelters open (84%), while 53% of the respondents did not drain the bathtub every three days. The majority of the respondents (46%) did not leave used goods in their yard; the respondents rarely used larvicide (85%) and maintained larva-eating fish (82%).

Table 1. Mosquito Breeding site Eradication (MBE) Behaviour in Tembalang sub district, Semarang.

Classification	Yes		No	
	f	%	f	%
Permanent bathtub use	70	70.0	30	30.0
Use of TPA cover	16	16.0	84	84.0
Drain the bath once every 3 days	47	47.0	53	53.0
Recycling used goods	46	46.0	54	54.0
Larvacide use	15	15.0	85	85.0
Culturing larvae eating fish	18	18.0	82	82.0

3.2 DHF incident in Tembalang sub district

Respondents of the study interviewed about the history of DHF showed 11% claimed to have had DHF. The data DHF cases got from the data report of health office of Semarang City. The number of DHF cases is reported more than the data from this survey. The DHF cases and the positive container distribution described in the villages in Tembalang sub district is presented in figure 1 and figure 2.

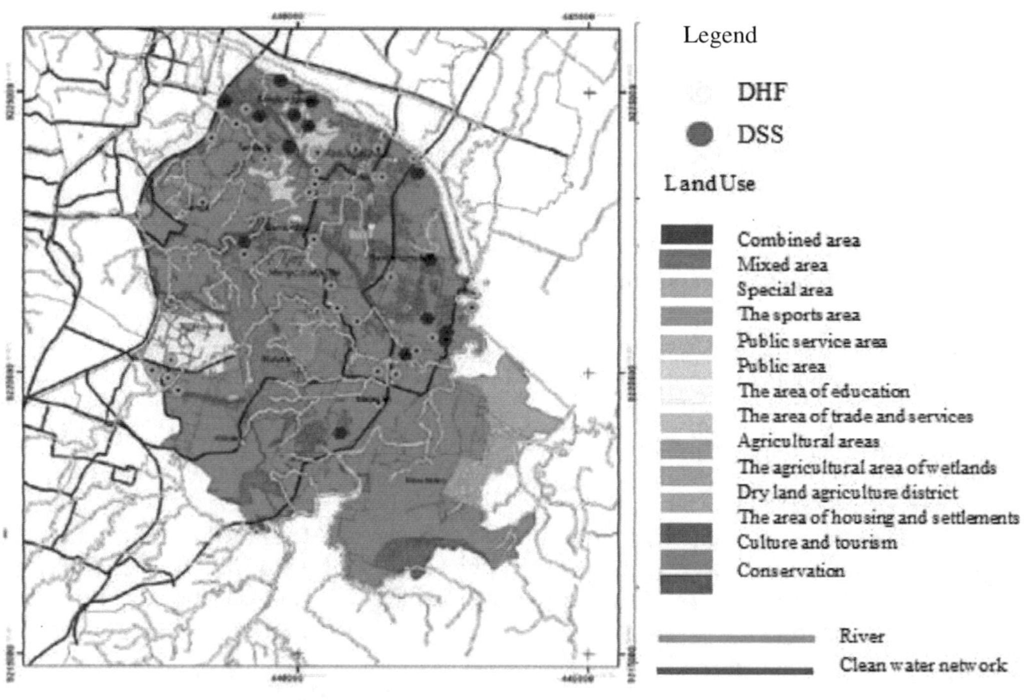

Figure 1. DHF distribution in all villages of Tembalang sub district.

Distribution DHF and DSS cases was found in Sendangguwo, Sendang mulyo, Kedungmundu, Meteseh and Tembalang villages (figure 1). They are an endemic with high cases of DHF and DSS. DHF cases tend to be found near the house with a positive larva (figure 2).

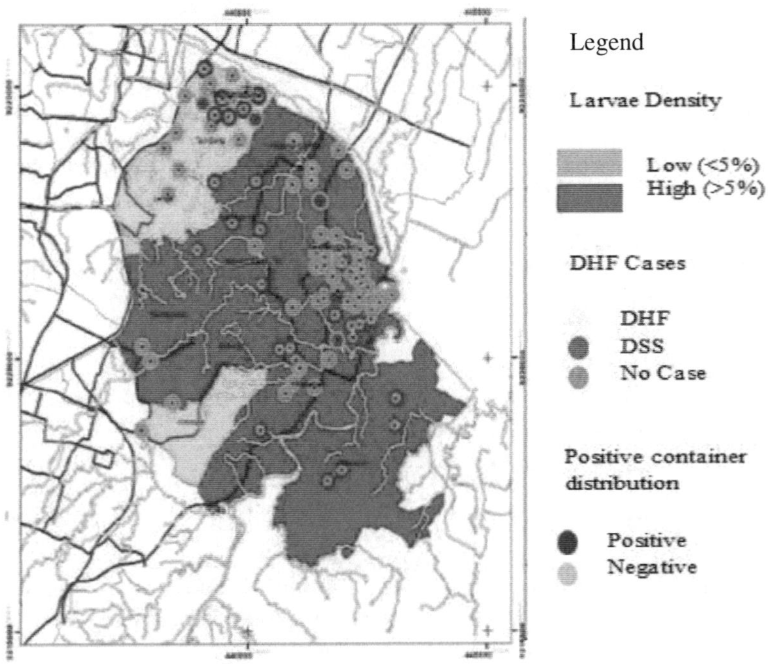

Figure 2. Positive container distribution in all villages.

One of the indicators of assessing success of vectors is by observing the density of larvae in the area whether or not the area is at risk for dengue transmission. Larvae density can be identified from Container Index (CI), House Index (HI), and Breteau Index (BI) in an area performing vector control [7].

Container Index (CI) describes the number of reservoirs found positive for larvae from the water reservoirs examined [10]. If the CI of an area is more than 5%, the area has a high risk [2]; whereas, if the CI is less than 5%, the area is safe from the transmission of dengue virus [11]. Various types of landfill found are likely to be used as a breeding place for DHF vectors. In this study, the CI of the 12 villages in Tembalang sub district was 9.83%, meaning that the vector density in the study area was quite high; while the type of the vector was *Ae. aegypti*.

Larvae density was closely related to the availability of breeding place, and *Ae. aegypti* preferred a container in the form of a water reservoir as the breeding place. Meanwhile, the availability of containers was closely associated with the behaviour of the community in collecting water. Nearly 83% of the people were categorized as having low MBE behaviour such as covering, washing the container, and recycling of household waste (called 3M plus on the local dengue program in Semarang). Many did not store water, close the water reservoir, and clean the container for at least 2 times per week.

Poor behaviour could worsen when the community increases; thus, a dense population environment greatly affected vector density in a region. Besides affecting vector density, poor behaviour and population density would also facilitate the transmission of DHF from one person to another vulnerable person. The higher population density and the closer the distance of houses, the more intensive the dengue disease spreads, especially in urban areas compared to rural areas [12]. Research finding of Prasetyowati showed that the IR (incidence Rate) of DHF in Bondowoso Regency with a population density of > 100,000 people/km^2 was still above the national standards. The dengue fever cluster spreading causes a densely populated area suffers higher possibility to be infected by DHF [12].

Prevention of dengue transmission has been carried out by breaking the chain of DHF transmission. Among the efforts are to eradicate mosquito breeding site (EMB) by means of "3M" plus, larvaciding, and fogging [2]. The EMB activities could be done by eliminating mosquito-breeding sites and recognizing mosquito bionomics in breeding. Semarang has dense population and suitable environment condition that encourages mosquito breeding easily [13].

The *Ae. aegypti* mosquitoes preferred to choose egg-laying places in a very specific habitat, which was influenced by container base material. The types of basic ingredients containers attracting the vector laid their eggs (oviposition) were cement, metal, soil, ceramics, and plastic as a breeding site of vectors. Cement based material is easily to be mossy, has rough surface and porous [14].This condition caused vector to prefer laying their eggs.

Physical environment such as temperature and humidity determined the life of *Ae. aegypti*. Tolerance to ambient temperature depends on mosquito species and geographical location such as tropical, sub-tropical, equator, and cold regions [5]. The optimal temperature for mosquito to grow and breed is in the range of 25^0C-35^0C [10]. The finding revealed that the air temperature in Tembalang sub-district is in the range of 27.8^0C-35^0C. In high humidity, mosquitoes generally live longer and spread faster, and they can survive at 60%-80% humidity [5]. Although the humidity in the study area was lower than 60%, *Ae. aegypti* identified adapted well to less than 60% humidity.

4. Conclusion

Ae. aegypti in Tembalang sub-district Semarang has a high capacity as a vector based on bionomic and biologic characteristics. Environmental factors, both related to the physical and social environment, strongly support the development of *Ae. aegypti*. Low community behaviour in eliminating mosquito breeding sites and temperature, humidity is a factor related to mosquito density in Tembalang District, Semarang. The high number of dengue cases in Tembalang District tends to be related to vector density based on the Container Index (CI) parameter.

5. References

[1] World Health Organization 2012 Report of a WHO Technical Working Group Meeting on Dengue Prevention and Control. (Geneva: WHO).

[2] Kementerian Kesehatan RI. 2010 Buletin Jendela Epidemiologi 2 1

[3] Tawastri A, Thavara U. 2010 Thailand : Biology and Ecology Section, Medical Entomology. (Thailand : National Institute of Health Departement of Medical Science, Ministry of Public Health).

[4] Bhattacharya S, Gething PW, Brady OJ, Messina JP, Farlow AW, Moyes, CL et al.2013 Natur 7446:504–507.

[5] Ridha MR, Rahayu N, Rosvita NA, Setyaningtyas DE 2013 *J Epidemiol dan Penyakit Bersumber Binatang*. 4:133–137.

[6] Dinas Kesehatan Kota Semarang 2015 Profil Kesehatan Kota Semarang Tahun 2015. (Semarang Dinas Kesehatan Kota)

[7] Dinas Kesehatan Kota Semarang. 2015. Rekapitulasi Kasus DBD 2006-2015. Semarang: Dinas Kesehatan Kota Semarang

[8] Koyadun S, Butraporn P, Kittayapong P 2012 *Interdisiplinary Perspectives on Infectious Diseases.*

[9] Al-Ghamdi K, Khan M A, Mahyoub J 2009 *Curr. World Environment.* 4 307-312 [10] Said GPS 2014 *J Kesehat Masy* 1326-337

[10] Aburas HM. 2007 *Inter. Scholarly and Sci. Research & Innovation* 1

[11] Service MW 1993 Mosquito Ecology Field Sampling Methods (London: Chapman & Hall)

[12] Prasetyowati I 2015 *Indones J Heal Sci.* 5.

[13] Martini M, Wahyuni CU, Subekti S, Notobroto HB, Hestiningsih R, Yuliawati, et al. 2017 *Adv Sci Lett*. 23 3367.

[14] Rahmawati I, Winita R. 2013. *J Kedokt*. 7.

Acknowledgements
We would like to extend our gratitude to the Dean of the Faculty of Public Health for the facilitation of funds and permits in carrying out research, as well as research respondents.

Identification of critical erosion prone areas in Temengor Reservoir Basin using Universal Soil Loss Equation (USLE) and Geographic Information System (GIS)

E Mohammad Basri[1], O Mohamad Adam[3], S Y Teh[2] and W O Wan Maznah[1,4,5*]

[1] School of Biological Sciences, Universiti Sains Malaysia, 11800 USM, Penang, Malaysia
[2] School of Mathematical Sciences, Universiti Sains Malaysia, 11800 USM, Penang, Malaysia
[3] Environmental Lab, Faculty of Engineering, University of Malaya, 50603 Kuala Lumpur, Malaysia
[4] Centre for Marine and Coastal Studies (CEMACS), Universiti Sains Malaysia, 11800 USM, Penang, Malaysia
[5] River Engineering and Urban Drainage Centre (REDAC), Universiti Sains Malaysia, Seri Ampangan, 14300 USM, Nibong Tebal, Penang, Malaysia

**Corresponding author: wmaznah@usm.my*

Abstract. In this study, the Geographic Information System (GIS) was integrated with the Universal Soil Loss Equation (USLE) model to identify the risk of erosion at 360,000 ha in the Temengor Reservoir Basin. GIS was utilized as a tool for generating, manipulating and spatializing data from government agencies for sediment yield modelling and offering spatial input data to the erosion model. Meanwhile, USLE was used to predict the spatial distribution of the sediment yield on the grid basis. The five main parameters used in this study were the rainfall erosivity factor (*R*), topographic factor (*LS*), soil erodibility factor (*K*), crop management factor (*C*) and practice support factor (*P*). The *R* factor was calculated based on the annual rainfall data of the study area. The soil survey data was used to generate the *K* value and Digital Improvement Model (DEM) of the study area was used to generate *LS* factor. The values of *C* and *P* factors were derived from the land use map. After generating all parameters, analysis was performed to estimate the soil erosion using USLE model with spatial information analysis approach. It was discovered that the average annual soil loss in the study area was 8 t ha^{-1} $year^{-1}$ and only 4% of the total area was under extreme erosion risk.

1. Introduction

Soil erosion is a major environmental global concern [1,2,3] which may cause sediment accumulation that will lead to intensification of water pollution problems [4,5]. Oldeman [6] reported that roughly 1094 Mha area is affected by soil erosion due to hydrological processes. The average rates of soil loss throughout the world as stated by Biggelaar et al. [7] are approximated between 12 and 15 t ha^{-1} $year^{-1}$, which mean, the land surface lost approximately 0.90–0.95 mm of soil every year [8]. Soil erosion is mostly affected by three major elements including water, wind and tillage [9]. According to the recent study, about 20–30 Gt of soil is being dislodged by water every year, while tillage erosion contributes 5 Gt per year [8]. Meanwhile, Lal [10] reported that there are 1094 million hectares of soil affected by soil erosion from water and 548 million hectares by wind.

The severity of soil erosion is mainly due to the variability in precipitation, soil characteristics, physiography, land use and human activities [11]. In Malaysia, soil erosion has become a major concern in recent years particularly in areas where land use is intensive for development, including urbanization and agricultural activities [12]. Invasion of development into environmentally sensitive areas has increase the number of soil erosion occasion, water pollution, sedimentation and subsequently flooding in downstream areas which also had significant impact on the peoples in and near the affected areas [12]. The effects of timber harvesting on soil erosion and sedimentation in Malaysia have been reported by several researchers including Salleh et al. [13], Baharuddin [14] and De Neergaard et al. [15]. Soil erosion not only affects the productivity of highland farmland but also water quality within catchment areas. [12]. This situation was created by the phenomenon of runoff on the surface of the vacant slope and half exposed to the river and eventually to the lake, which will certainly reduce the depth of the lake in the long run [16]. The introduction of chemicals from pesticides and fertilizers as a result of agricultural activities has increased the concentration of excess nutrients such as nitrogen and phosphorus along with the heavy metal content in water and lake sediment [12] which subsequently lead to eutrophication problem in reservoirs [17].

There are many tools that researchers have introduced to estimate soil erosion including empirical and process-based models, for example the Water and Soil Assessment Tool (SWAT) [18], the Water Erosion Prediction Project (WEPP) [19], the Universal Soil Loss Equation (USLE) [20] and others. Process-based models, such as WEPP and SWAT typically require intensive calibration and multiple input parameters with extensive field data collection [21]. Furthermore, the variability of input parameters, leading to uncertainties, may be the main reason why process-based erosion models usually cannot produce as good an empirical model [22]. Thus, empirical models such as USLE is commonly used to estimate the soil erosion by water worldwide [23,24,3] including in Malaysia [25,26], because of its simplicity, and understandable from a functional perspective [21]. The USLE method predicts the average annual long-term erosion rate on the field by considering data from rainfall patterns, soil types, topography, crop systems and management practices [3].

The Temengor Lake catchment area has experienced rapid economic growth over the last decade. Land use activities around Temengor Lake, especially in the south, have changed from primary forest to logging area either for timber production or for agricultural purposes [27,28]. This type of developmental activities has a substantial influence on soil erosion and affect the ecological, biological and hydrological functions of the lake system. In this study, the USLE model was used to estimate the potential soil loss and develop an erosion risk maps for Temengor Reservoir catchment areas with the help of GIS that can be used to implement best management practices in future soil conservation programs.

2. Materials and methods

2.1 Study area

This study was conducted in the entire area of Temengor Reservoir Basin (5°56'4.12"N, 101°34'53.55"E and 3°40'37.19"N, 101°31'13.29"E) of Gerik in the northern state of Perak (figure 1) which includes approximately 360,000 ha of tropical rainforest with elevations ranging from 130–1500 m above sea level. Temengor Reservoir (15, 734 ha) which located in the middle of the whole basin has an initial storage capacity up to 6.05 x 10^9 m^3 with mean depth of 40 m. This reservoir is surrounded by three forest reserves namely Belum Forest Reserve (137,167 ha), Temengor Forest Reserve (148,870 ha) and Gerik Forest Reserve (37,220 ha) which serve as a water catchment area. Among them, only Belum Forest Reserve is considered a fully protected forest whereas others are still open for logging activities [27,28]. In general, the Temengor Reservoir Basin can be described as having a typical tropical monsoon climate with relatively high temperatures (24°C - 30°C) and humidity between 70-98%. It receives high rainfall in April and October each year reaching 3,000 mm per year at times while experiencing during low rainfall in February and July [29]. The major land use for this area is forestry which constitutes about 66.26% of the entire catchment area followed by road and river (21.64%) and rubber plantation (10.9%) with a projected human population of 616 by 2015 [30].

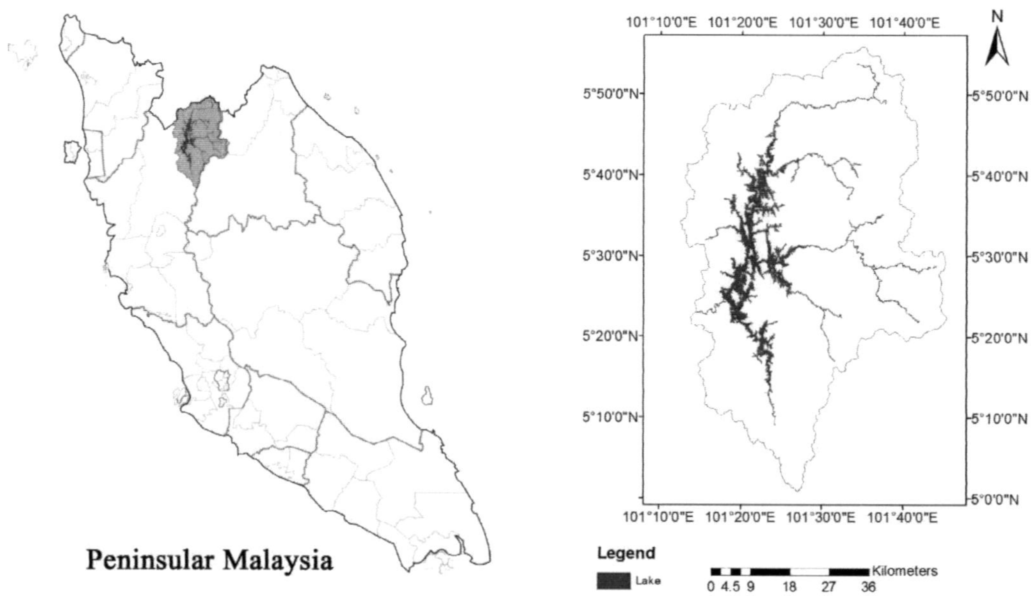

Figure 1. Study area of Temengor Reservoir, Perak

2.2 The Universal Soil Loss Equation (USLE)

In the present study, USLE was applied in the Temengor Reservoir Basin in the spatial domain using GIS. All USLE factors were obtained as raster geographic layers (grid) after processing the original data. The cell size of 100m × 100m was considered as the basic operating unit for soil erosion analysis as proposed by [31]. Then, all factors were combined to calculate the final erosion risk map (figure 2). Soil erosion can be classified into five different levels as listed in table 1.

The equation used was:

$$A = R \times K \times LS \times C \times P$$

Where:

> *A* is the average annual soil loss (tons ha^{-1} $year^{-1}$),
> *R* is the rainfall erosivity (MJmm ha^{-1} h year),
> *K* is the soil erodibility factor (tons ha^{-1} R $unit^{-1}$),
> *LS* is the topographic factor,
> *C* is the cropping management factors, and
> *P* is the practice support factor

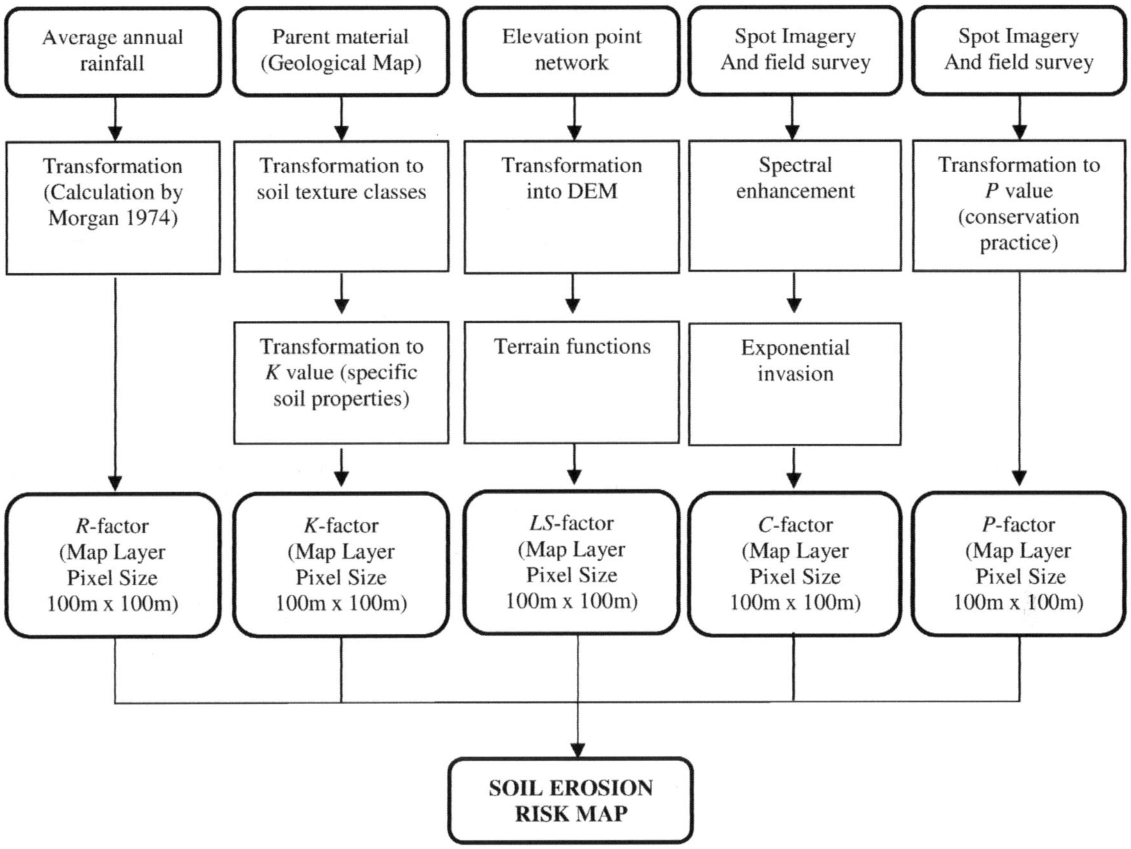

Figure 2. Procedural Flow Chart of this study

Table 1. Soil loss classification [3]

Level	Soil loss (ton ha^{-1} $year^{-1}$)
Very low erosion	0 - 1
Low erosion	1 - 5
Medium erosion	5- 10
High erosion	10 - 50
Extreme erosion	> 50

2.2.1 Rainfall erosivity factor (R). Rainfall erosivity factor (*R*) plays the most vital factor in USLE compared to the other input parameters [32]. Rainfall erosivity (*R*) was defined as the product of the total kinetic energy multiplied by a maximum intensity of 30 minutes of rain [33]. Many methods have been introduced by researchers to calculate the annual rainfall erosivity factor [35,1]. For Peninsular Malaysia, Morgan [36] developed an equation to calculate the *R* factor based on annual precipitation by analysing the rainfall data obtained from the Department of Irrigation and Drainage of Malaysia (DID) and Tenaga Nasional Berhad (TNB) from 2007 to 2017.

The equation used was:

$$R = \frac{(9.28P - 8838.15)}{100}$$

Where *P* is the mean annual rainfall (mm)

2.2.2 Soil erodibility factor (K). The soil erodibility factor (*K*) defines the resistance of the soil to both detachment and transport [35]. This factor illustrates the impact of soil properties and soil profile characteristics on soil loss. All equations for measuring *K* factor are associated to soil texture, soil organic matter, and percentage of sand, silt, and clay in soil [37]. In Malaysia, *K* factor value was already determined by Department of Irrigation and Drainage of Malaysia (DID) according to the type of soil. In this study, the main soil type for the whole catchment area is 'Steepland' with a clay soil structure.

2.2.3 Topographic factor (LS). Topographic factor (*LS*) is the slope length gradient factor consisting of slope length (*L*), and slope steepness (*S*) [38]. The rate of soil erosion by water is greatly influenced by both slope length and slope steepness based on its gradient and slope percentage [35]. The *LS* factor can be calculated using the methodology proposed by Wischmeier [39]:

$$LS = \left(\frac{\lambda}{\psi}\right)^m \times (0.065 + 0.046S + 0.0065S^2)$$

Where:

λ	= sheet flow path length (m)
Ψ	= 22.13 for SI Units and 72.6 for English Units (BU)
S	= average slope gradient (%)
m	= 0.2 for $s < 1$,
	= 0.3 for $1 \leq s < 3$,
	= 0.4 for $3 \leq s < 5$,
	= 0.5 for $5 \leq s < 12$ and
	= 0.6 for $s \geq 12\%$

Currently, GIS is considered an important tool for the study of environmental management. Therefore, almost all researchers use Digital Elevation Model (DEM) and GIS tools to measure *LS* in soil erosion studies [38,40]. In the present study, the *LS* factor was calculated using ArcGIS and DEM model obtained from a 10-meter interval contour with a 1:50,000 scale based on topographic maps from the Department of National Mapping, Malaysia (JUPEM).

2.2.4 Cropping management factor (C). The value of *C* factor ranges between 1 and 0. *C* equals to 1 indicating no cover exist and surface is considered to be barren soil, while *C* approaching zero shows very strong protective effects [3]. The *C* factor can be calculated based on current land use and crop management in the area [41]. In the present study, the *C* factor was determined using a land use map provided by the Department of Agriculture of Malaysia (DOA). The Department of Irrigation and Drainage of Malaysia (DID) summarizes the list of *C* factor values for several crops and conditions available in Malaysia; for agricultural and urbanization areas (table 2) and for unaffected and undisturbed land (table 3).

2.2.5 Support practice index (P). *P* factor is defined as the effect of land use or agricultural system on soil erosion. The *P* factor can affect the potential erosion of water runoff by applying contour effects, crop rotation, and terraced contour farming [42,43]. If no erosion control solution exists, then the value of *P* is 1. In this study, due to the lack of data distributed spatially for the *P* factor, it was set to 1 for the entire study area, assuming that no protection measures were taken [23,44].

Table 2. Cropping management factor (*C*) for agricultural and urbanized areas [24]

Erosion control treatment	C factor
Mining areas	1.00
Agricultural areas	
Agricultural crop	0.38
Horticulture	0.25
Cocoa	0.20
Coconut	0.20
Oil palm	0.20
Rubber	0.20
Paddy (with water)	0.01
Urbanized areas	
Residential	
Low density (50% green area)	0.25
Medium density (25% green area)	0.15
High density (5% green area)	0.05
Commercial, Educational and Industrial	
Low density (50% green area)	0.25
Medium density (25% green area)	0.15
High density (5% green area)	0.05
Impervious (Parking lot, road, etc.)	0.01

Note: average runoff condition

Table 3. Cropping management factor (*C*) for forested and undisturbed lands [24]

Erosion control treatment	C factor
Rangeland	0.23
Forest/Tree	
25% cover	0.42
50% cover	0.39
75% cover	0.36
100% cover	0.03
Bushes/Scrub	
25% cover	0.40
50% cover	0.35
75% cover	0.30
100% cover	0.03
Grassland (100% coverage)	0.03
Swamps/mangrove	0.01
Water body	0.01

Note: average runoff condition

3. Results and discussion

The average annual rainfall from 2007 to 2017 was used to determine the *R* factor for the whole catchment area using the equation described in section 2.2.1. The mean annual rainfall ranged from 1606 to 2219 mm resulting in rainfall erosivity of 327 to 815 mm, with the highest *R* factor recorded in the middle of the Temengor catchment area (figure 3a). As mentioned earlier, the equation for measuring *K* factor were related to soil texture, soil organic matter, and percentage of clay, sand and mud in the soil [37]. Therefore, *K* is one of the most arduous factors, which requires considerable time, cost and resources for field observations and analysis. In Malaysia, the value of *K* factor has been determined by the Department of Irrigation and Drainage Malaysia (DID) by type of soil. Based on the data obtained from the DID, the main soil type for the entire catchment area was 'Steepland' with clay structure and the *K* value for soil type and structure was 0.06 (figure 3b). Slope length and slope steepness play a vital role in the rate of soil erosion by water in terms of its gradient and slope percentage [35]. However, longer slopes do not necessarily correspond to higher soil loss irrespective of the three-dimensional complexity of the terrain [3]. Thus, many researchers come to agreement that the rate of soil lost hinge on the three-dimensional distribution of the area [45,46]. In this study, the combined spatial distribution of *LS* factor was obtained from the DEM of the study area. The *LS* factor in the Temengor Reservoir Basin ranging from 0 to 67.3 (figure 3c) with *LS* value less than 5 was 72.6 % (265,146 ha) while *LS* value exceeding 5 was 27.4% (100,069 ha). The potential for erosion is higher in areas with high *LS* values due to the increased slope angle and the slope length [45]. Areas with *LS* value of more than 5 are more likely to experience soil erosion than areas with slopes of less than 5 [47]. High LS values will lead to very strong rainfall rates and makes the soil erosion worse. The *C* factor in the study area was in between 0.01 to 0.45 with the highest value of *C* focused in the southern Temengor River Basin where most of the logging and rubber plantation activities were located (Figure 3d). The *P* factor was set to 1 for the whole study area (figure 3e) due to the lack of data distributed spatially which means no protection measures were taken [23,44].

Average annual soil erosion (A) was calculated by multiplying the raster data generated from each of the USLE factors as described in section 2.2. The average forecasts of soil loss in the study area ranged from "very low erosion" to "extreme erosion" (table 4; figure 3f). The data shows that throughout the whole Temengor Reservoir Basin, the area where the erosion is high, occupies 28.8% of the total area especially at high elevations zone. The "extreme erosion" area accounted for 4.0% of the study area mostly recorded in the southeastern Temengor Reservoir Basin where logging activity was located. The north-east and south-western basins have less erosion rates than other areas.

In this study, the level of erosion in the Temengor Lake basin, largely influenced by the combined effects of *R*, *LS* and *C* factors. *LS* and *C* factors were also giving a huge impact on soil erosion on Siruvani River Basin as noted by Thomas & Thrivikramji [11]. Meanwhile, Magesh and Chandrasekar [48] identified *R* and *LS* factors as the leading factors affecting the soil erosion in the sub-basin of the Tamiraparani River. Many studies [49,50,51] have discovered that soil loss rates in catchment areas are more sensitive to water runoff (*R* factor). From the *R* factor map (figure 3a), the lowest value of R was recorded in the southwestern Temengor Reservoir Basin where low erosion was detected. Various researchers [11, 48] indicated that *LS* factor plays a significant role in soil loss rate and is the most challenging factor in the USLE model [24]. Similar result was shown in this study where the soil erosion was high, occupying 28.8% of the total area, mostly occurred at high elevations area. The *C* factor was closely linked to land use in the area. Most researches have concluded that land use influenced both results and characteristics of soil loss [52,53,54]. Different land uses will give a different levels of soil erosion. Pham [55] proposes that plantation forests will erode the most. In studies conducted by several researchers [56,57,58], it was found that land use for agricultural land resulted in the worst erosion. In this study, the agricultural land together with logging areas were severely eroded. When natural plants are cleared for agricultural activities, exposed soil layers are often blown away by wind or washed away by rain compared to forests filled with vegetation canopy [2].

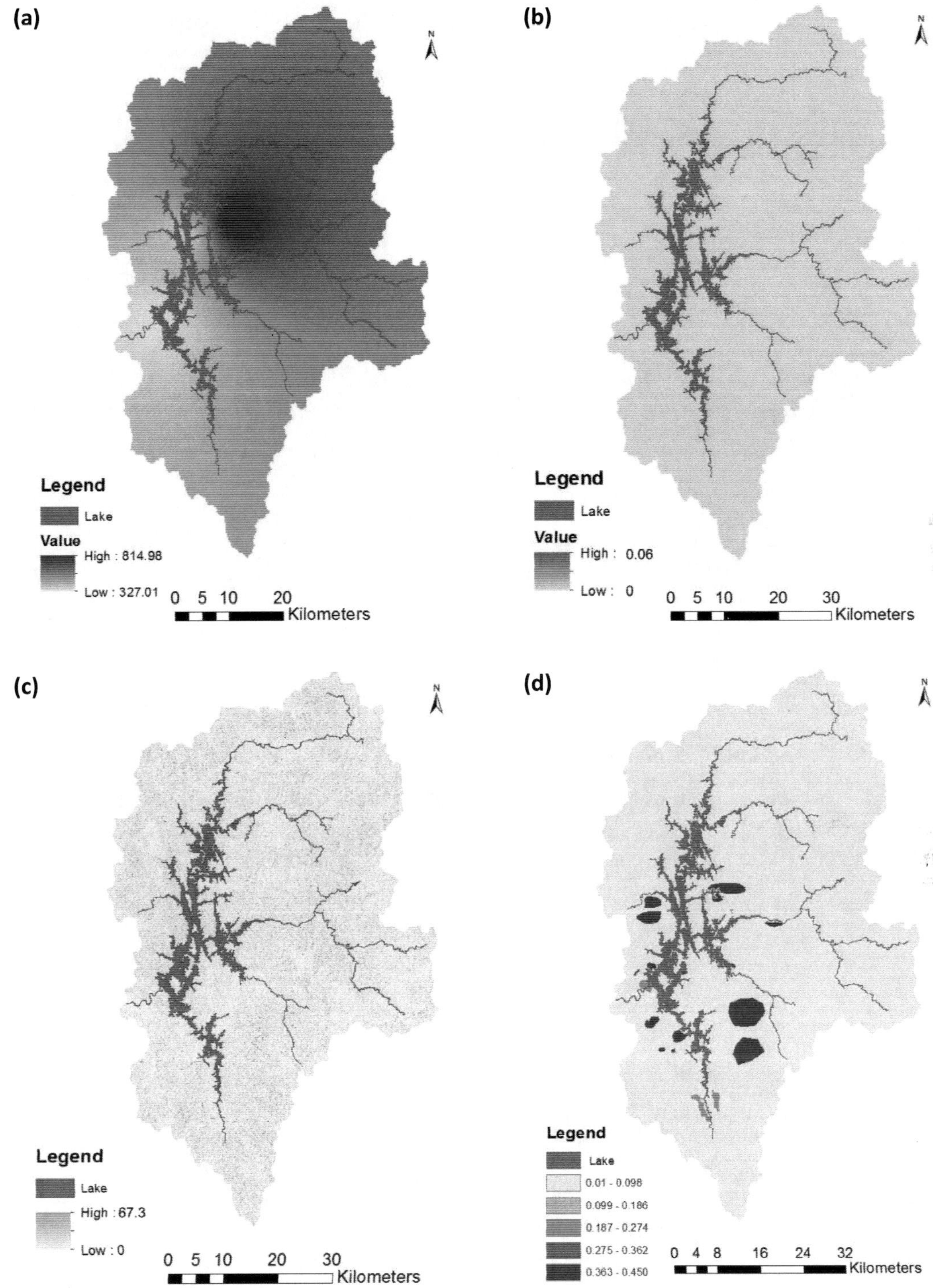
(a)
Legend
Lake
Value
High : 814.98
Low : 327.01
0 5 10 20 Kilometers
(b)
Legend
Lake
Value
High : 0.06
Low : 0
0 5 10 20 30 Kilometers
(c)
Legend
Lake
High : 67.3
Low : 0
0 5 10 20 30 Kilometers
(d)
Legend
Lake
0.01 - 0.098
0.099 - 0.186
0.187 - 0.274
0.275 - 0.362
0.363 - 0.450
0 4 8 16 24 32 Kilometers

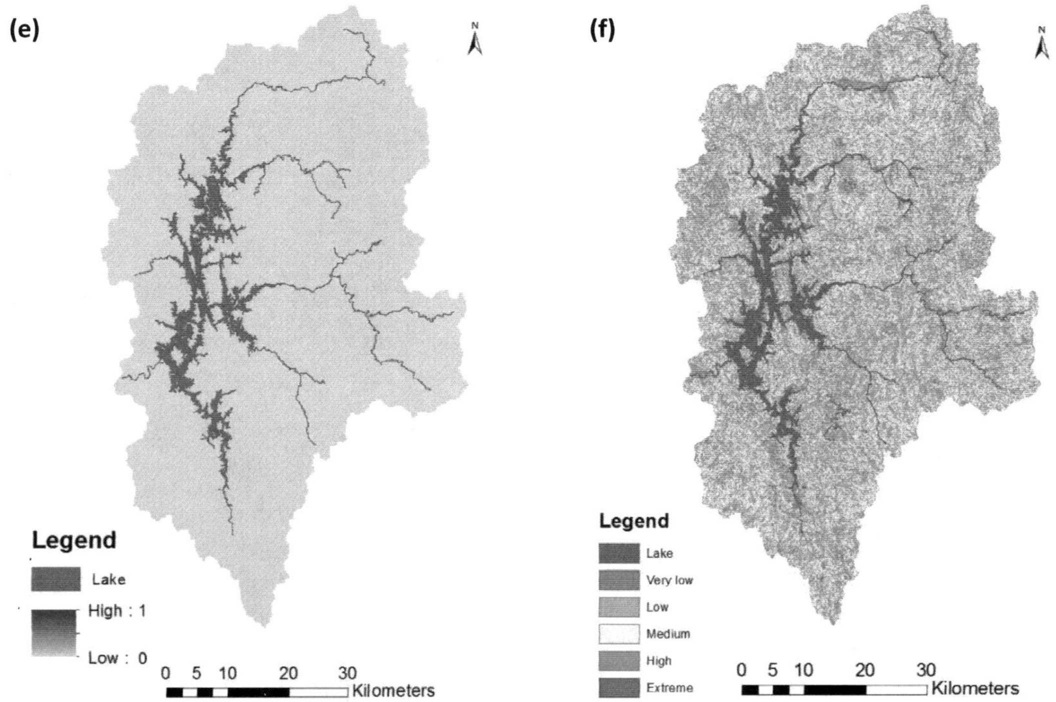

Figure 3. a. Rainfall erosivity factor (*R*), b. Soil erodibility factor (*K*), c. Topographic factor (*LS*), d. Cropping management factor (*C*) and e. Support practices factor (*P*), f. Soil erosion

Table 4. Soil erosion rate in Temengor Reservoir Basin

Erosion level	Soil loss (tons ha^{-1} $year^{-1}$)	Area (ha)	Percentage of total area (%)
Very low	0 - 1	34,330	9.4
Low	1 - 5	41,635	11.4
Medium	5- 10	153,756	42.1
High	10 - 50	105,182	28.8
Extreme	> 50	14,609	4.0
Water		15,703	4.3
Total		365,215	100

4. Conclusion

The average soil loss rate in the Temengor Reservoir Basin was around 8 tons ha^{-1} $year^{-1}$. 42.1% of the Temengor Reservoir Basin area represents moderate erosion risk while high erosion risk (28.8%) was located in the southeast which includes timber and some agricultural areas. The effects of parameters affecting soil loss in the USLE model indicate that the rainfall erosivity factor (*R*), topographic factor (*LS*) and crop management factor (*C*) have the greatest impact on soil loss in the study area. This study also demonstrates that the USLE soil erosion model combined with GIS is an efficient tool for handling large volumes of data needed for soil loss studies in the catchment area.

5. References

[1] Parveen R & Kumar U 2012 Integrated approach of universal soil loss equation (USLE) and geographical information system (GIS) for soil loss risk assessment in upper south Koel Basin, Jharkhand *Journal of Geographic Information System* **4**(6) 588–596

[2] Ganasri B P & Ramesh H 2016 Geoscience Frontiers Assessment of soil erosion by RUSLE model using remote sensing and GIS - A case study of Nethravathi Basin. *Geoscience Frontiers* **7**(6) 953–961

[3] Pham T G, Degener J & Kappas M 2018 Integrated universal soil loss equation (USLE) and Geographical Information System (GIS) for soil erosion estimation in A Sap basin: Central Vietnam *International Soil and Water Conservation Research* **6**(2) 99–110

[4] Abidin R Z 2017 Erosion risk assessment: A case study of the Langat River bank in Malaysia. *Int. Soil Water Conserv. Res.* **5** (1) 26-35

[5] Suhaimi H, Jamal M & Ahmad A 2018 Assessment of river bank erosion at Kilim River, Langkawi using geospatial technique *IOP Conf. Series: Earth and Environmental Science* **169**(2018) 1–10

[6] Oldeman L R 1994 An international methodology for an assessment of soil degradation, land georeferenced soils and terrain database, In the collection and analysis of land degradation data. *FAO-RAPA Publication 1994/3 Bangkok* pp. 35-60

[7] Biggelaar C D, Lal R, Wiebe K. & Breneman V 2003 The global impact of soil erosion on productivity: I: Absolute and relative erosion-induced yield losses. *Journal of Advances in Agronomy* **81** 1-48

[8] FAO & ITPS 2015 *Status of the world's soil resources (SWSR)* – Main Report Food and Agriculture Organization of the United Nations and Intergovernmental Technical Panel on Soil (Rome, Italy) p 608

[9] Ghabbour E A, Davies G, Misiewicz T, Alami R A, Askounis E M, Cuozzo N P, Filice A J, Haskell J M, Moy A K, Roach A C & Shade J 2017 National Comparison of the Total and Sequestered Organic Matter Contents of Conventional and Organic Farm Soils. *Advances in Agronomy* 146 1-35

[10] Lal R 2001 Soil degradation by erosion *Journal of Land Degradation & Development* **12**(6) 519-539

[11] Thomas J Joseph S & Thrivikramji K P 2018 Estimation of soil erosion in a rain shadow river basin in the southern Western Ghats, India using RUSLE and transport limited sediment delivery function *International Soil and Water Conservation Research* **6**(2) 111–122

[12] Mir S I, Sahid I, Gasim M B & Rahim S A 2015. Prediction of Soil and Nutrient Losses from the Lake Chini. *Journal of Physical Science* **26**(1) 53–70

[13] Salleh M N, Abdul Rahim N & I. Mushrifah 1981 Logging and water quality. A Malaysian experience Paper presented at *International MAB Symposium on Man's Impact on Ecosystem Dynamics-with Emphasis on influences on Environmental Pollution, Tokyo, Japan*

[14] Baharuddin K 1988 Effect of logging on sediments yield in a hill dipterocarp forest in Peninsular Malaysia *J. Tropic. Forest Sci* **1**(1) 56-66

[15] De Neergaard A, Magid J and Mertz O 2008. Soil erosion from shifting cultivation and other smallholder land use in Sarawak, Malaysia *Agric. Eco. Environ.* 125 182-190

[16] Mir S I, Gasim M B, Ismail B S, Sahibin A R & Ekhwan T M (2012). Estimation of the Rate of Soil Erosion in the Tasik Chini Catchment, Malaysia Using the RUSLE Model Integrated with the GIS. *Australian Journal of Basic and Applied Sciences* **6**(12) 286–296

[17] Sharip Z & Zakaria S 2008. Lakes and Reservoir in Malaysia: Management and Research Challenges *Proceedings of Taal2007: The 12th World Lake Conference* 1349–1355

[18] Arnold J G, Srinivasan R R, Muttiah S & Williams J R 1998. Large-area hydrologic modelling and assessment: Part I. Model development. *J. American Water Resour. Assoc.* **34**(1) 73-89

[19] Flanagan D C & Nearing M A (eds.) 1995 USDA-Water Erosion Prediction Project (WEPP)

Hillslope Profile and Watershed Model Documentation. NSERL Report No. 10, National Soil Erosion Research Laboratory, USDA-Agricultural Research Service, West Lafayette, Indiana.
[20] Wischmeier W H & Smith D D 1965 Predicting rainfall-erosion losses from cropland east of Rocky Mountains: guide for selection of practices for soil and water conservation. Agricultural handbook 282, US Department of Agriculture.
[21] Zhu M (2015) Soil erosion assessment using USLE in the GIS environment: a case study in the Danjiangkou Reservoir Region, China. *Environmental Earth Sciences* **73**(12) 7899–7908
[22] Jetten V, Govers G & Hessel R 2003 Erosion models: quality of spatial predictions. *Hydrol Process* 17 887–900
[23] Sotiropoulou A, Alexandridis T, Bilas G & Karapetsas N 2011 A user friendly GIS model for the estimation of erosion risk in agricultural land using the USLE *International Conference on Information and Communication Technologies* 795–801
[24] Devatha C P & Renukaprasad M S 2015 Estimation of Soil loss using USLE model for Kulhan Watershed, Chattisgarh- A case study In *Aquatic Procedia* **4**(4)1429–1436
[25] Elsheikh F R A, Ouerghi S, Elhag A R, Sulayman A & Arabia S 2015 Soil erosion risk map based on Geographic Information System and Universal Soil Loss Equation (USLE) Case Study: Terengganu. *Ind. J. Sci. Res. and Tech.* **3**(2) 38–43
[26] Pradhan B, Chaudhari A & Buchroithner J A M F 2012 Soil erosion assessment and its correlation with landslide events using remote sensing data and GIS: a case study at Penang Island, Malaysia **184**(2) 715–727
[27] Abdullah A R, Weng C N & Som A P M 2011 The Potentials and Perils of Ecotourism in Belum Temengor Forest Complex. *World Applied Sciences Journal* (12) 1–9
[28] Kaur R, Teresa O, Kim C L & Chin A Y 2011 A Survey on Mass Movements of the Vulnerable Plain-Pouched Hornbill in the Belum-Temengor Forest Complex. *The Raffles Bulletin of Zoology* (24) 171–176
[29] Aiman Hanis J, Abu Hassan A, Nurita A T & Che Salmah M R 2014. Community structure of termites in a hill dipterocarp forest of Belum- Temengor Forest Complex, Malaysia: Emergence of pest species Community structure of termites in a hill dipterocarp forest of Belum – Temengor Forest Complex, Malaysia: emergence of. *The Raffles Bulletin of Zoology* (62) 3–11
[30] Gerik Government Web Portal 2018. *Places of interest: Banding Island, Perak (Tourism Section).* http://www.gerik.gov.my/ms/pelawat/destinasi-menarik/pulau-banding. Online Access [15 July 2018]
[31] Singh G & Panda R K 2017 Grid-cell based assessment of soil erosion potential for identification of critical erosion prone areas using USLE, GIS and remote sensing: A case study in the Kapgari watershed, India. *International Soil and Water Conservation Research* **5**(3) 202–211
[32] Jebari S 2009 *Water erosion modelling using fractal rainfall disaggregation – A study in semiarid Tunisia* (Sweden: Water resources engineering Lund University) p 135
[33] Wischmeier W H & Smith D D 1978 *Predicting rainfall erosion losses—a guide to conservation planning*. U.S. Department of Agriculture (Agriculture Handbook No. 537)
[34] Gitas I Z, Douros K, Minakou C, Silleos G N & Karydas C G 2009 Multi- temporal soil erosion risk assessment in N. Chalkidiki using a modified USLE raster model. *EARSel eproceedings* 8 40–52
[35] DID 2010 *Guideline for erosion and sediment control in Malaysia*. (Malaysia: Department of Irrigation and Drainage) 125p
[36] Morgan RPC 1974 Estimating regional variations in soil erosion hazard in Peninsular Malaysia *Malayan Nature Journal* 28 94-106
[37] Millward A A & Mersey J E 1999 Adapting the RUSLE to model soil erosion potential in a mountainous tropical watershed *Journal of Catena* **38**(2) 109–129
[38] Panagos P, Borrelli P & Meusburger K 2015 A new European slope length and Steepness factor (LS-factor) for modelling soil erosion by water *Journal of Geosciences* (5) 117-126

[39] Wischmeier W H 1975 Estimating the Soil Loss Equation's Cover and Management Factor for Undisturbed Areas, In Present and Prospective Technology for Predicting Sediment Yields and Sources *Proceedings Sediment-Yields Workshop* (U.S. Department)
[40] Remortel R D V, Hamilton M E & Hickey R J 2001 Estimating the LS factor for RUSLE through iterative slope length processing of digital elevation data within arclnfo grid *Journal of Cartography* **30**(1)27-35
[41] Stone R P & Hillborn D 2012 *Universal soil loss equation (USLE)* Factsheet Order No.12-051 pp1-8
[42] Kim S H & Julien P Y 2006 Soil erosion modelling using RUSLE and GIS on the MIHA Watershed *Water Engineering Research Journal* **7**(1) 29-41
[43] Kuok K K K, Mah D Y S & Chiu P C 2013 Evaluation of C and P factors in universal soil loss equation on trapping sediment: Case study of Santubong River *Journal of Water Resource and Protection* (5) 1149-1154
[44] Erdogan E H, Erpul G & Bayramin I 2007 Use of USLE/GIS methodology for predicting soil loss in a semiarid agricultural watershed. *Environmental Monitoring and Assessment* **131**(1–3) 153–161
[45] Mitasova H, Hofierka J, Zlocha M & Iverson L R 1996 Modelling topographic potential for erosion and deposition using GIS *Journal of Geographical Information Systems* **10**(5) 629–641
[46] Moore I D &Wilson J P 1992 Length-slope factors for the revised universal soil loss equation: Simplified method of estimation *Journal of Soil and Water Conservation* **47**(5) 423-428
[47] Lihan T, Khodri N F, Mustapha M A, Rahman Z A & Idris W M R 2018. Potensi Hakisan Tanih di Lembangan Sungai Bilut, Raub, Pahang menggunakan Integrasi. *Sains Malaysiana* **47**(10) 2241–2249
[48] Magesh N S & Chandrasekar N 2016 Assessment of soil erosion and sediment yield in the Tamiraparani sub-basin, South India, using an automated RUSLE-SY model *Environmental Earth Sciences* **75**(16) 1-17
[49] Jain S K, Kumar S & Varghese J 2001 Estimation of soil erosion for a Himalayan watershed using GIS technique *Water Resources Management* **15**(1) 41–54
[50] Kayet N, Pathak K, Chakrabarty A & Sahoo S 2018 Evaluation of soil loss estimation using the RUSLE model and SCS-CN method in hillslope mining areas. *International Soil and Water Conservation Research* **6**(1) 31–42
[51] Dabral P P, Baithuri N & Pandey A 2008 Soil erosion assessment in a hilly catchment of North Eastern India using USLE, GIS and remote sensing *Water Resources Management*, **22**(12) 1783-1798
[52] Fu B J, Wang J F, Lu Y H, He C S, Chen L D & Song C J 2009. The effects of land-use combinations on soil erosion: A case study in the Loess Plateau of China. *Progress in Physical Geography* **33**(6) 793-804
[53] Garcia R & Jose M 2010 The effects of land uses on soil erosion in Spain: A review. *Catena Journal* **81**(1) 1-11
[54] Wang G, Jiang H, Xu Z, Wang L & Yue W 2012 Evaluating the effect of land use changes on soil erosion and sediment yield using a grid-based distributed modelling approach. *Journal of Hydrological Process* **26**(23) 3579–3592
[55] Pham H T 2008 Soil erosion risk modeling within upland landscapes using remotely sensed data and the RUSLE model (A case study in Huong Tra district, Thua Thien Hue province, Vietnam) *International Symposium on Geoinformatics for Spatial Infrastructure Development in Earth and Allied Sciences*
[56] Erskine W D, Mahmoudzadeh A & Myers C 2002 Land use effects on sediment yields and soil loss rates in small basins of Triassic sandstone near Sydney, NSW. *Australia Journal of Catena* **49**(4) 271-287
[57] Gangcai L, Jianhui Z, Guanglong T & Chaofu W 2005 The effects of land uses on purplish soil

erosion in hilly area of Sichuan Province, China. *Journal of Mountain Science* **2**(1) 68-75
[58] Tran T P, Chau V T T, Nguyen B N & Huynh V C 2014 Modelling soil erosion within small mountainous watershed in central Vietnam using GIS and SWAT *Journal of Resources and Environment* **4**(3) 139-147

Acknowledgements
This study was supported by the Research University Grant Scheme (1001/PBIOLOGI/870039) from Universiti Sains Malaysia.

IOP Conf. Series: Earth and Environmental Science **380** (2019) 012012 doi:10.1088/1755-1315/380/1/012012

Population structure and gonadal maturity stage of endemic and alien fish dominant species in Lake Matano, South Sulawesi

SH Nasution[1*] and R Dina[1]

[1]Research Centre for Limnology, Indonesian Institute of Sciences
Jl. Raya Jakarta Bogor CSC-BC, Km 46, Cibinong 16911, Indonesia

**Corresponding author: syahroma@limnologi.lipi.go.id*

Abstract. Lake Matano, one of the lakes in the Malili Lake complex, is located in East Luwu District, South Sulawesi Province. Lake Matano is an ancient lake that is very deep with high biota and endemicity, including fish. The dominant endemic fish found on Lake Matano is *Telmatherina antoniae*. At present,dominant alien fish which is Louhan fish is also found at this site. This paper aims to convey the latest information regarding population structure of the dominant endemic and alien fish in Lake Matano,namely relative abundance distribution, size structure (length and weight), growth patterns and gonadal maturity stages. Sampling was carried out in May and August 2016 at five stations. Fish were caught using gill nets with seven mesh sizes and installed for at least two hours.The relative abundance of fish at each station is determined based on the average of catch per unit effort, namely the number of fish caught per hour of gillnet installation(ind/hour), with growth pattern determined based on the length weight relationship, and the stage of gonadal maturity determined morphologically.*Telmatherina antoniae* and Louhan were found at all stations with the highest relative abundance in MT4 (Petea) and MT2 (Nuha), respectively. The length of *T.antoaniae* ranged from 64 to 105 mm (82.24 ± 0.61 mm) and weight 2.3 to 11 grams (5.62 ± 0.14 mm). The length of Louhan ranged from 51 to 185 mm (96.42 ± 3.36 mm) and weight from 2.5 to 102.6 grams (21.66 ± 2.36 grams). In May and August both *T.antoniae* and Louhan were found with gonadal maturity stage I to IV.

1. Introduction

Lake Matano, located in the Soroako area (south Sulawesi), is an ancient lake which is a tectonic lake, as a part of the Lake Malili complex (Matano, Mahalona, Towuti, Lantoa, and Masapi). Lake Matano is located in the upper stream, Lake Mahalona in the middle stream and Lake Towuti in the lower stream. Lake Matano has an area of 164 km^2 with oligotrophic trophic status, water visibility up to 20-23 meters, and a depth of 596 m.Lake Matano is utilized by the surrounding community as a recreation area, cage culture for fish, a source of hydroelectric power, source of water for domestic needs, and furthermore for nickel mining activity around Lake Matano area.

Indonesia has 7,000-8,500 species of fish, and 3,000 of them are freshwater fish, most of which are ornamental fish. There are 14 species of fish endemic to Lake Matano [1].Lake Matano has endemic biota such as Opudi fish (*Telmatherina* sp.) which consists of nine species and currently extending to 10 species [2]. Moreover, butini (*Glossogobius matanensis*) is other endemic fish whose presence is

IOP Conf. Series: Earth and Environmental Science **380** (2019) 012012 doi:10.1088/1755-1315/380/1/012012

scarce due to anthropogenic activities and the presence of alien species such as the Louhan fish. In addition to fish species, there are also types of endemic shrimp, crabs and mollusks in this lake.

Fish diversity in Indonesia currently faces threats from various human activities that can cause a decline in the fish diversity. Of the 87 species of Indonesian fish that are threatened with extinction, 66 species (75%) are known as freshwater fish [3]. Most (68%) of these endangered freshwater fish are endemic fish [15]. According to [4], destruction of freshwater fish stock is mostly as a consequence of habitat damage (35%), introduction of alien species (30%), over exploitation (4%), pollution and global warming. As an example, some activities aroundLake Towuti have the potential to damage the water environment such as: 1) legal and illegal deforestation; 2) industrial sawmill yielding waste of saw-mill; 3) industrial nickel mining yielding waste; and 4) intensive fishing [5]. It is estimated that alien fish species entered Lake Matano would threaten the existence of endemic fish (*Telmatherina* spp.) which inhabits Lake Matano.

This paper aims to convey the latest information regarding population structure of the dominant endemic and alien fish species in Lake Matano, namely relative abundance distribution, size structure (length and weight), growth patterns and Gonadal Maturity Stages (GMS).

2. Research Methodology

2.1. Study Site

This research was conducted at Lake Matano, East Luwu Regency, South Sulawesi in May and August 2016 which represented the rainy and dry season. Fish were caught at five representative observation stations in littoral zones of Lake Matano (figure 1). Descriptions of the sampling stations are presented in table 1.

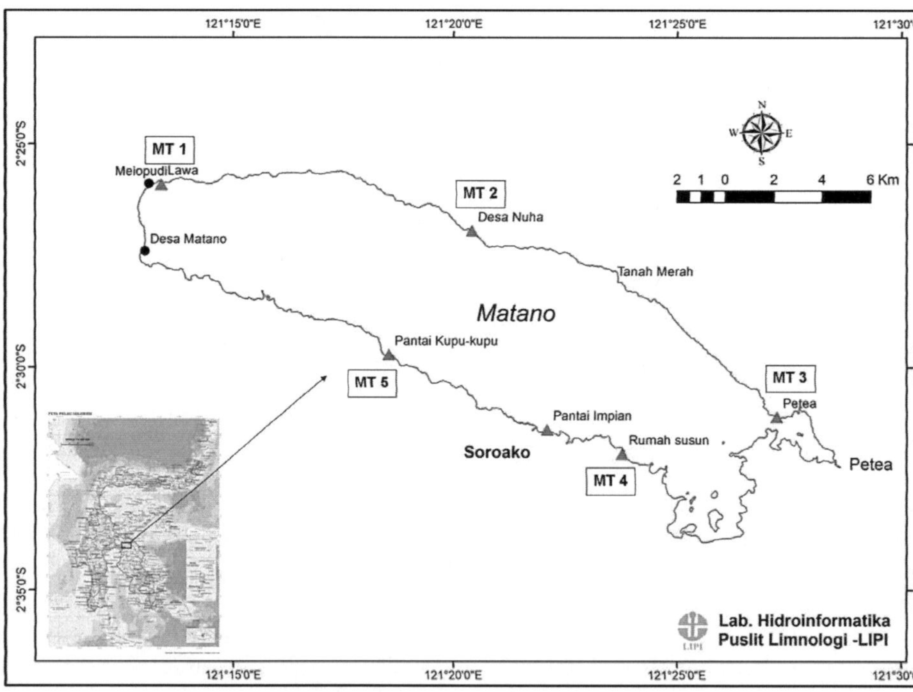

Figure 1. Sampling stations in Lake Matano (MT.1-MT.5).

Table 1. Descriptions of the sampling stations.

No.	Sampling Stations	Description	Figure of station
1	MT1 LawaRiver	Lawa River is an inlet of Lake Matano, while around the sampling location there is still a forest, the substrate is gravel to a depth of 2 meters and at depths of 5 and 10 meters dominated by sand, the disturbance from anthropogenic activities (residential settlements) is very minimal.	
2	MT2 Nuha	Location is close to Nuha Village, there are fishing villages, boat docks, boat crossings from Nuha Village to Sorowako. Around the village there is still natural vegetation cover (forest), the type of substrate is small gravel to a depth of 2 meters and at depths of 5 and 10 meters is sand.	
3	MT3 Petea	Outlet of lake Matano is Petea station, a more open area, there is a former forest clearing to be converted into a plantation area. Formerly a former area of nickel tailings waste disposal, the substrate from a depth of 0 to 2 meters is mud and fine sand and at a depth of 5-10 meters is sand. There are swamps.	
4	MT4	There are cage aquaculture ,in front of flats, no natural vegetation, near PT Vale, land clearing for pepper plantations, ,the substrate from a depth of 0 to 2 meters is gravel and sand and at a depth of 5-10 meters is sand.	
5	MT5 Butterflay beach/ Bubble beach	There are Butterfly group around this area, also known as Bubble beach, gas bubbles which are suspected to be H_2S are found in the waters, the riparian zone around the location is still dominated by native vegetation, the substrate at depth of 0-2 meters is dominated by gravel and at depth of 5 and 10 meters is sand.	

2.2. Type of Research
The research method used is descriptive quantitative approach. The research station was determined based on different habitat types representing the waters of Lake Matano.

2.3. Research Focus
This paper aims to convey the latest information regarding population structure of the dominant endemic and alien fishes in Lake Matano, in the form of relative abundance distribution, size structure (length and weight), growth patterns and the gonadal maturity stages (GMS)

2.4. Techniques of Data Collection and Data Analysis
Fish samples were caught using experimental gill nets consisting of seven different mesh sizes, namely: ¾, 1, 1½, 2, 2½, 3, and 3½ inches with lengths of 25 m each and 1.8 m high, so that the total length of one unit is 175 m.The net were equipped with buoys at the top and ballasts at the bottom. Thenetwas installed at each station from the direction of the beach towards the free waterin the upper water column [6] for two hours during the day between 08.00-12.00 pm (figure 2).The number of catch at each observation station was recorded and separated bysex. Then the total length and total wet weight were measured.

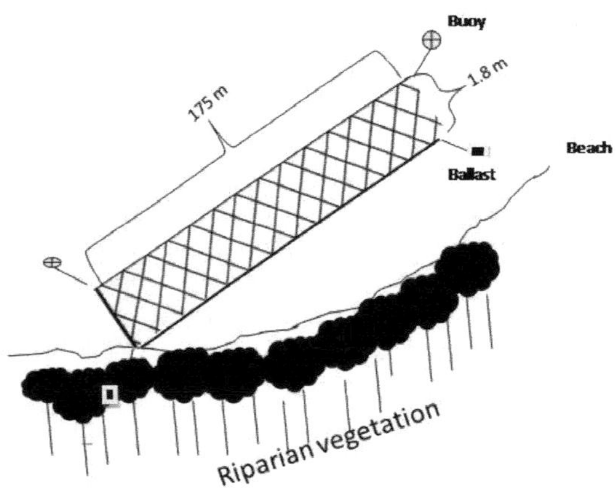

Figure 2. Position of fish sampling using Gill net on Lake Matano.

Total length is the length of the fish measured from the anterior tip of the longest jaw to the most posterior part of the caudal fin. The total length can be measured to the nearest unit below it [7] or to the nearest unit [8]. The total length was measured to the nearest unit below it. The weight of the fish measured was the total wet weight. Total wet weight is the total weight of the fish's body tissue and the water contained in it. Total wet weight measurements were carried out in the field [9]. A measuring board (accuracy of 1 mm) was used to measure length and the digital scales of ACIS BC 500 series (accuracy of 0.1 g) to measure the weight of fish. Another tool used was a surgical device to determine the Gonadal Maturity Stage (GMS). Furthermore, the GMS was determined morphologically by following Cassie's classification [10] and [11]. For identification purpose, some of the fish samples were preserved in 4% formaldehyde solution and then immersed in 70% alcohol solution. Fish were identified using references from [12], [13], [14], and [15] at the laboratory.

Data were analysed descriptively and analytically. The relative abundance of fish at each observation station is seen based on the average value of CPUE (individual/hour), namely the number of fish caught per hour per the length of installation of gillnet. Length, weight, and GMS of fish were analysed using descriptive statistics. Fish growth patterns were determined using a lengthweight relationship equation, $W = aL^b$ [16]. If $b = 3$, then the growth pattern is isometric (length increase is proportional to weight gain). If $b \neq 3$ then the growth pattern is allometric (length increase is not proportional to weight gain). If $b > 3$, then the growth pattern is a positive allometric where weight gain is more dominant than the increase in length, whereas if $b < 3$, then the growth pattern is negative allometric where length is more dominant than weight gain [17]. To test the null hypothesis that $\beta = \beta 0$ can be calculated t. If the value of t> t (α / 2, n-2) then the null hypothesis is rejected and if t <t (α / 2, n-2) the null hypothesis fails to be rejected [18] in this case $\beta 0 = 3$. Some of physical and chemical water quality parameters (temperature, pH, and dissolved oxygen/DO) was measured directly using Horiba U-51 Water Quality Checker / WQC Horiba U-51. This measurement was carried out together with the gill net installation.

3. Results and discussion

3.1. Environment

In general, the entire range of water quality in research stations still meet the requirements for fisheries activities according to the quality standards of the Ministry of Environment [19] as shown in table 2. The water quality standard tolerance threshold for fluctuating temperature parameters does not exceed 3°C, pH 6-9 and minimum dissolved oxygen (DO) 3 mg/L. The surface water temperature of Lake Matano during the study ranged from 26.9 to 29.08 °C.

Acidity and alkalinity of a lake is measured with a pH value, on a scale of 1-14. pH is defined as the negative log of hydrogen ion concentration. The concentration of hydrogen ions controls the chemical condition of the nutrients and carbon dioxide in the waters. Carbon dioxide in the waters is used by plants for photosynthesis [15]. In general, lake pH ranges from 6-9, Lake Matano has a neutral pH to close to bases. Carbon dioxide is free to maintain the equilibrium of HCO_3 - in solution and is called CO_2 balance. Based on observations (the water pH of Lake Matano ranges from 8.39 - 8.87), it can be predicted that the equilibrium of the reaction is in the form of carbonate.

Distribution of dissolved oxygen (DO) in water is very important in the metabolic processes of organisms. Solubility of some inorganic nutrients is affected by DO. Changes in nutrient distribution will affect the speed of growth of microorganisms. The dissolved oxygen value in Lake Matano ranged from 5.5 - 7.67 mg/L.

Table 2. Water quality of Lake Matano.

No.	Stations	Temperature (°C)	pH	DO (mg/L)
1	MT1(Lawa River)	27.5-28.5	8.39-8.65	5.50-7.73
2	MT2(Nuha)	26.9-28.8	8.75-8.87	7.26-7.63
3	MT3 (Petea)	27.9-28.3	8.74-8.79	7.29-7.46
4	MT4 (Front of flats)	28.7-29.08	8.47-8.64	7.29-7.54
5	MT5 (Butterfly beach)	28.3-28.9	8.39-8.70	6.12-7.67

3.2. Catch composition

The dominant endemic fish species found in May and August was *Telmatherina antoniae* by 57% and 32%, while the dominant alien fish species was Louhan, by 27% and 42% (figure 3). Species of *Telmatherina* sp. belongs to the Telmatherinidae family, which is an endemic tribe in Sulawesi waters, estuary waters and mangrove swamps in Papua New Guinea [15].

Furthermore [15] explained that the Telmatherinidae family such as *Telmatherina antoniae* (figure 4a) had two dorsal fins where the first dorsal fin had simple spines and generally had a colour difference between males and females. Louhan fish found in Lake Matano is thought to be the result of artificial cross hybrid of *Amphilopus citrinellus* and *Cichlasoma trimaculatum* (figure 4b). The percentage of Louhan has increased even to become the dominant type as a whole. Furthermore, both *Telmatherina antoniae* and Louhan will be analysed in this paper.

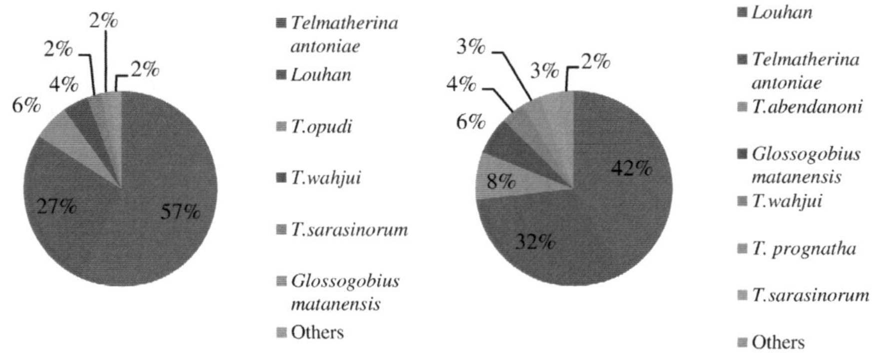

(a) (b)

Figure 3. Catch composition on (a) May and (b) August 2016 in Lake Matano.

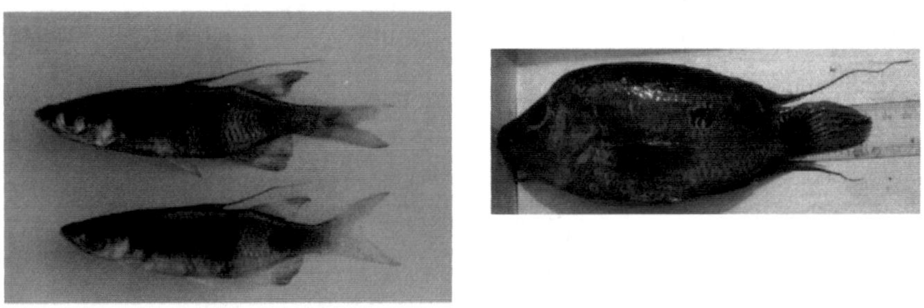

(a) (b)

Figure 4. Dominant endemic fish species (a) *Telmatherina antoniae* and dominant alien fish species (b) Louhan.

3.3. Dominant Fish Population Structure

3.3.1. Relative abundance distribution.

The distribution of relative abundance of *T. antoniae* at each sampling location can be seen in table 3 and figure 5.

Table 3. Catch per unit effort of *Telmatherina antoniae* in each location.

Stations	May	August	Average
MT 1(Lawa River)	5	10.5	7.75
MT 2 (Nuha)	6.5	1	3.75
MT 3 (Petea)	64	5.5	34.75
MT 4 (Front of flats)	19	0.5	9.75
MT 5 (Butterfly beach)	1	2.5	1.75

Telmatherina antoniae found in all sampling stations with a relative abundance score in MT3 (Petea) is very high compared to other stations. [21] also reported that *T.antoniae* was found in nine observation sites on Lake Matano with abundance in Petea is low. The habitat of *Telmatherina* fish has diverse substrates. As reported by [22] that *T.antoniae* spawned on habitats with substrate sand/mud, cobblestone to rocks, and aquatic plants. *Telmatherina antoniae* spawn at a depth of less than 1 m to more than 10 m. All types of *Telmatherina* fish are substrate spawner which are types of fish that lay and spawn their eggs on the substrate.

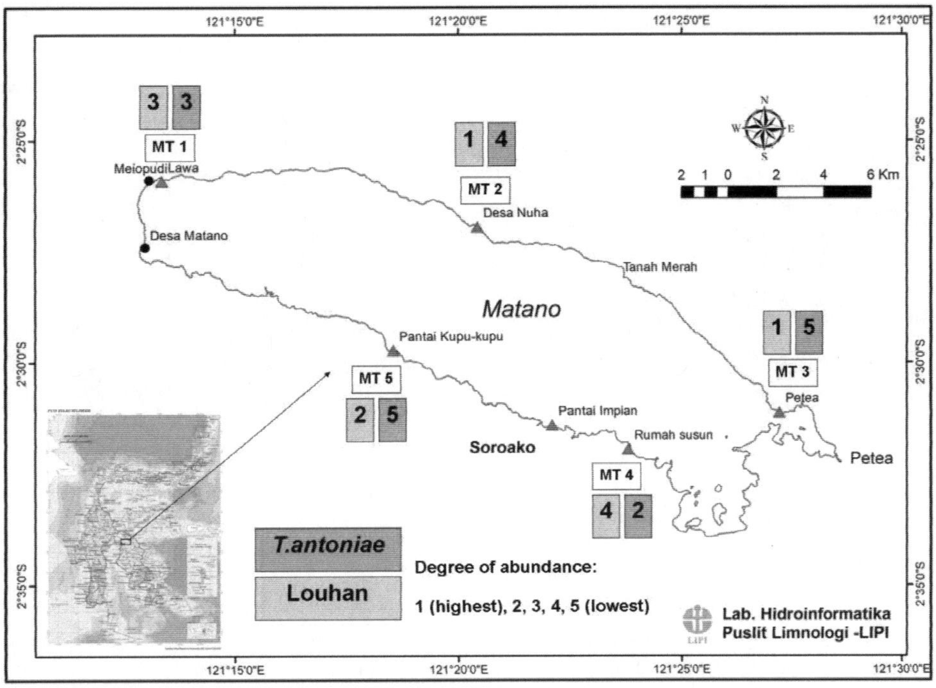

Figure 5. Relative abundance distribution of *T.antoniae* and Louhan in Lake Matano.

As previously stated, Louhan fish is a very dominant alien fish in Lake Matano and its population continues to increase. The distribution of relative abundance of Louhan fish in Lake Matano can be seen in table 4 and figure 5.

Table 4. Catch per unit effort of Louhan in each location.

Stations	May	August	Average
MT 1(Lawa River)	6.5	6.5	6.5
MT 2 (Nuha)	29	4	16.5
MT 3 (Petea)	0.5	3.5	2
MT 4 (Front of flats)	2	2.5	2.25
MT 5 (Butterfly beach)	8.5	10.5	9.5

Louhan are found in all sampling stations with the highest relative abundance at MT2 (Nuha) stations followed by MT5 (Butterfly Beach/Bubble beach) and in other stations with relatively low abundance. As reported by [23] that Louhan fish occupy a wider habitat from the edge of the lake to a depth of more than 13 m and adult fish generally prefer habitat with rocky substrate.

3.3.2. Size Distribution

- *Telmatherina antoniae*

Based on figure 6 it is known that the length of female *T.antoniae* ranged from 69-105 mm with the highest frequency in the class interval 79 to 83 mm. Whereas male fish were found with a length range of 64 to 102 mm with the highest frequency in the class interval of 79 to 83 mm. The size of both female and male fish was relatively smaller than previously reported by [24] in [21], namely 64 to 106 and 64 to 120 mm, respectively female and male fish. Spatially, there is a tendency that fish length at MT4 and MT5 stations was greater than in other locations. The weight ranged from 3.2 to 11 gram for female and 2.3 to 11 gram for male (figure 7).

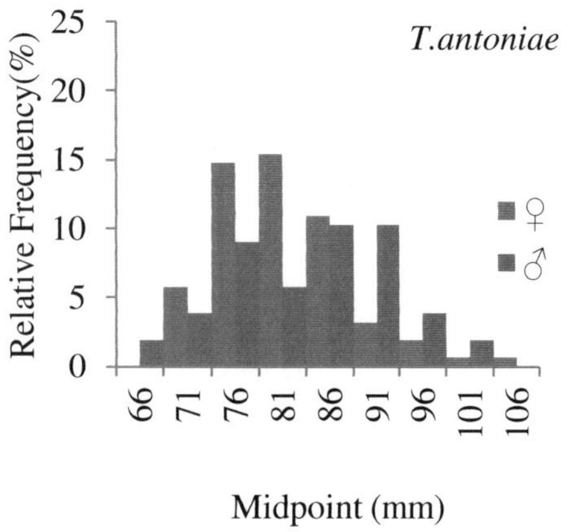

Figure 6. Total length distribution of *T.antoniae* in Lake Matano.

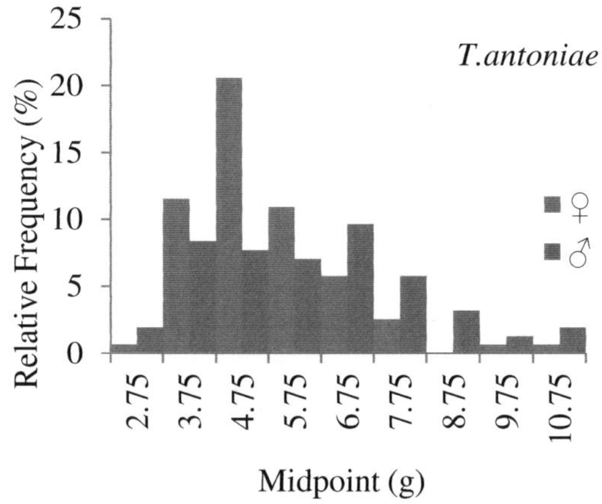

Figure 7. Wet weight size distribution of *T.antoniae* in Lake Matano.

- Louhan

The length and weight of analysed Louhan male fish varied more than female fish, as shown in figure 8 and 9. Based on these figures, it was found that the length and weight of male fish were in the class 59-187 mm and 8.75-96.95 grams. The size of female fish is also smaller than male fish, namely the length in the class interval 59-171 mm and the weight of 8.75-84.35 grams. The Louhan fish analysed were smaller than the Louhan fish size reported by [25] which was around 250 mm. This is not much different from the information posted on the website www.fishbase.org which reports that the maximum size of *Amphilopus citrinellus* fish is 244 mm SL and *Chiclosoma trimaculatum* 365 mm SL.

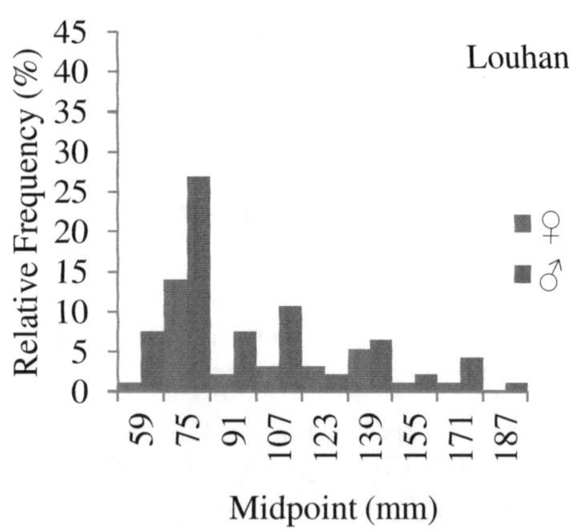

Figure 8. Total length of Louhan fish in Lake Matano.

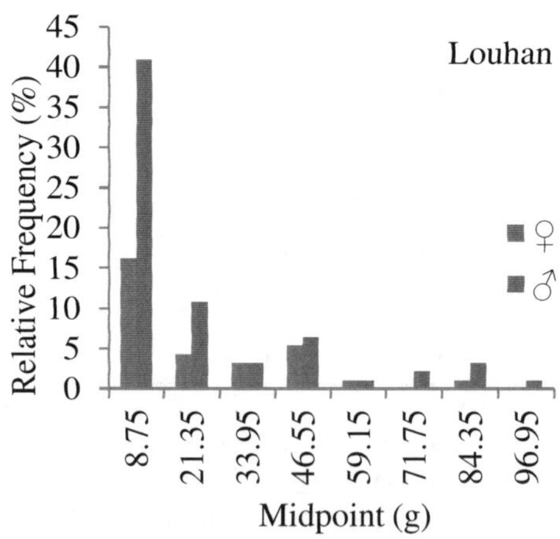

Figure 9. Wet weight of Louhan fish in Lake Matano.

3.3.3. Length-weight relationship.

Length-weight relationship models of male, female and overall *T.antoniae* are $W = 6E\text{-}06L^{3.12}$, $W = 5E\text{-}05L^{2.63}$ and $W = 1E\text{-}05L^{2.95}$ (figure 10). The exponent value of b for length and weight in individual male is 3.12 and female is 2.63. T-test (α=0.05) results on the regression coefficient value (b) shows the value of $b < 3$ for female and b=3 for male and all fish. This value indicates that the growth pattern of female is allometric, both male and overall *T.antoniae* is isometric. This means that the growth of *T.antoniae* weight is balanced with length growth. [26] states that the length weight relationship will slightly deviate from the cubic law ($b \neq 3$) because the environment and fish's condition often changes. Biologically the value of b relates to the condition of the fish, while the condition of the fish depends on the food, age, sex and maturity of the gonads [10].

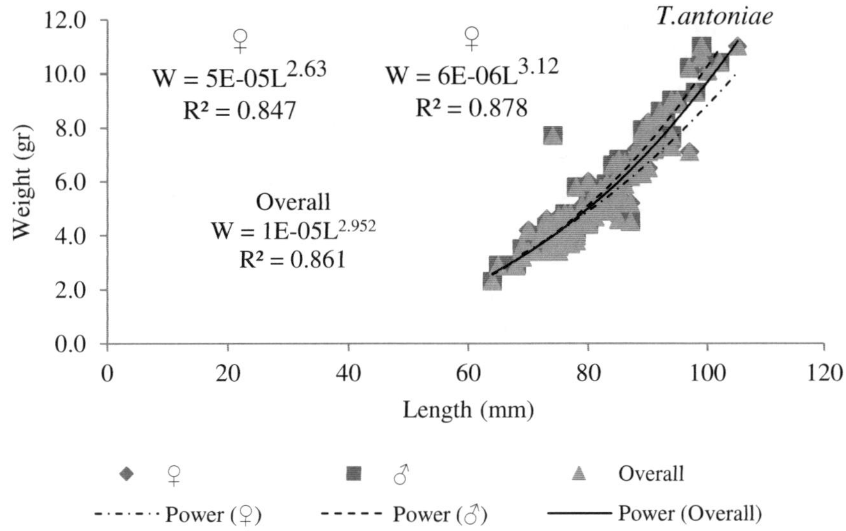

Figure 10. Length-weight relationship of *T.antoniae* in Lake Matano.

The length weight relationship model of the male, female, and overall Louhan are $W = 2E\text{-}05L^{3.02}$, $W = 9E\text{-}06^{3.15}$, and $W = 1E\text{-}05L^{3.06}$ (figure 11). T-test results show that b value of female is higher than 3 and b value for male and overall of Louhan is equal to 3. This value indicates that the growth pattern of female is allometric, male and overall Louhan fish is isometric which means that fish weight growth proportionally with its length.

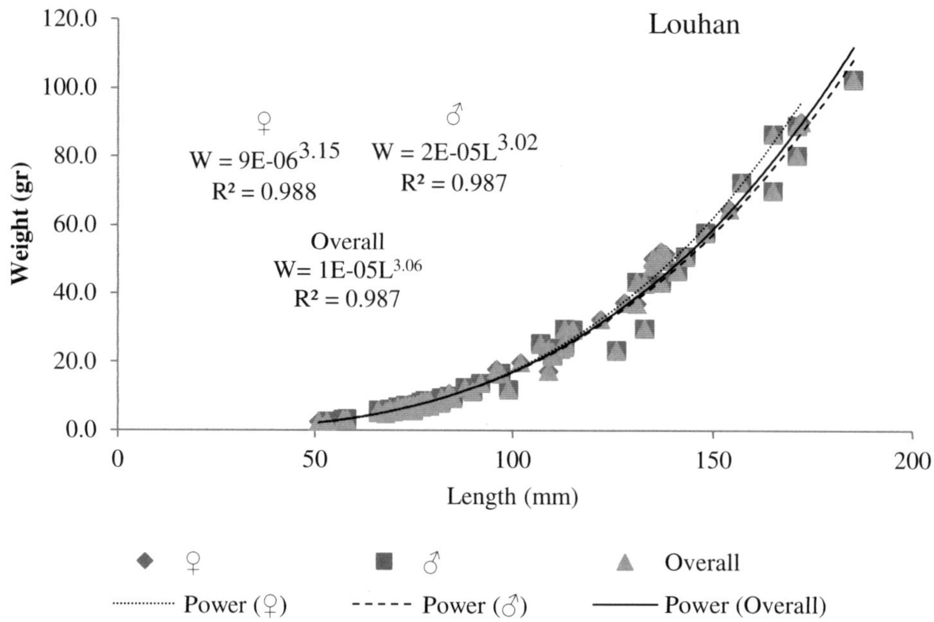

Figure 11. Length-weight relationship of Louhan in Lake Matano.

3.3.4. The gonadal maturity stage (GMS) of the dominant fish

- *Telmatherina antoniae*

According to [27] that the testicular and ovarian organs in most teleostei fish are a pair of organs located in the body cavity. But in some types of fish, testicular and ovarian pairs develop only one organ. Bonti-bonti fish have one reproductive organ that developed. The same thing according to [24] found in opudi fish (*T. antoniae*) originating from Lake Matano, [28] in rainbow selebensis (*T. celebensis*) from Lake Towuti, and [29] reported on fish beseng-beseng (*T. ladigesi*) from several rivers in Maros, South Sulawesi. These fish are classified as sexual dimorphism, which means that the fish has properties that can be used to distinguish male and female fish.

Gonadal maturity stage (GMS) is a certain stage of gonadal development before and after the fish spawns. The development of gonads is part of the reproductive process of fish. The occurrence of gonadal development as a result of the vitellogenesis process which is the process of deposition of egg yolk on each egg cell. Before spawning occurs, most of the fish's metabolic products are used for gonad development purposes.

Gonadal development of *T.antoniae* is classified into five stages, namely GMS I (undeveloped), II (early development), III (developing), IV (mature), and V (post spawning) both in male and female fish that refer in [30], [31], [32] and [33] with modifications (figures 12 and 13).

The GMS of *T.antoniae* fish in May and August was I - IV, but male *T.antoniae* in August at MT4 and MT5 stations was not found. In August more male and female *T.antoniae* were found at the GMS IV. This shows that *T.antoniae* fish are more likely to spawn, especially at MT1 station so that they can recruit (figures 12 and 13).

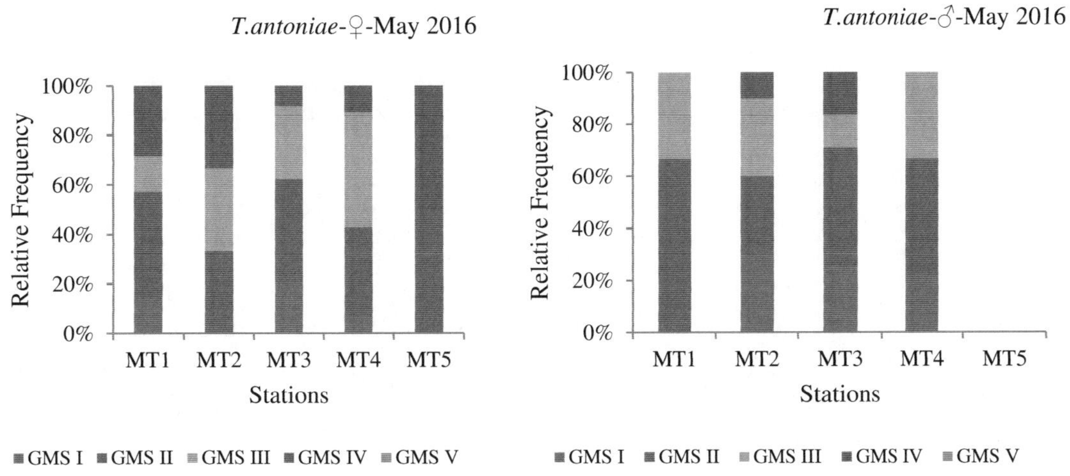

(a) (b)

Figure 12. Gonadal maturity stage of *T.antoniae* (a) female and (b) male in May at Lake Matano.

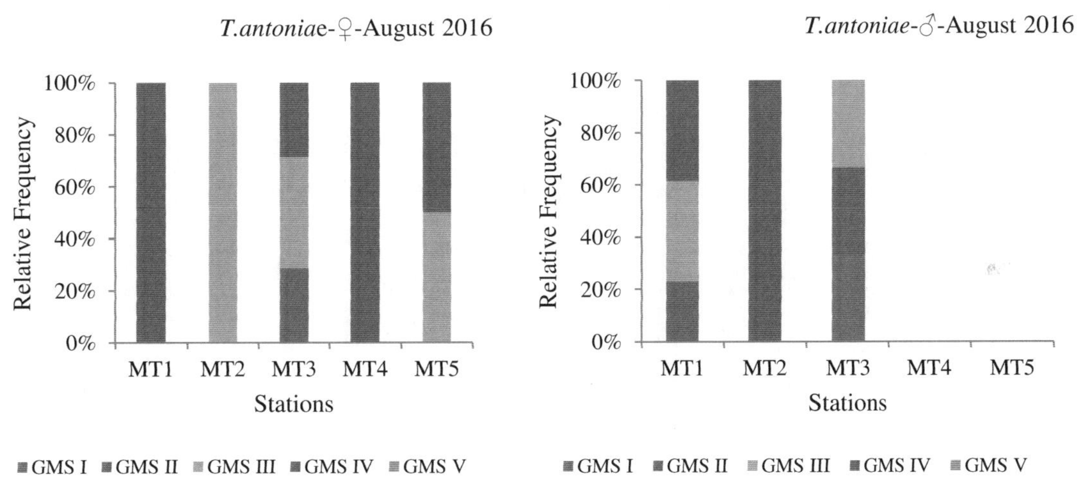

(a) (b)

Figure 13. Gonadal maturity stage of *T.antoniae* (a) female and (b) male in August at Lake Matano.

- Louhan

During May and August 2016, GMS of Louhan ranged from I to IV (figures 14 and 15) at all stations, except at MT3 and MT4 stations female fish was not found on May. This allows Louhan fish to reproduce quickly. Louhan fish also eat eggs of endemic fish that are attached to the substrate [23], so that it is estimated that the recruitment of endemic fish will not be achieved. Some of these facts indicate the potential threat of Louhan fish to the Lake Matano ecosystem. However, dominant endemic fish such as opudi (*T. antoniae*) with gonad maturity stage (GMS) I-IV was also found in almost all stations. Therefore it is expected that *T. antoniae* fish can reproduce and recruitment will be achieved, so that the population could be sustainable.

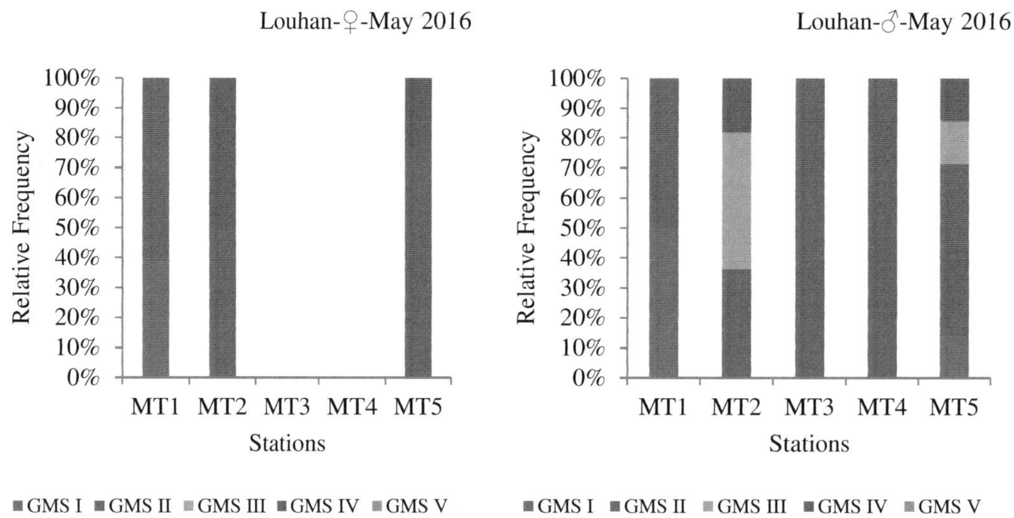

(a) (b)

Figure 14. Gonadal maturity stage of Louhan (a) female and (b) male in May at Lake Matano.

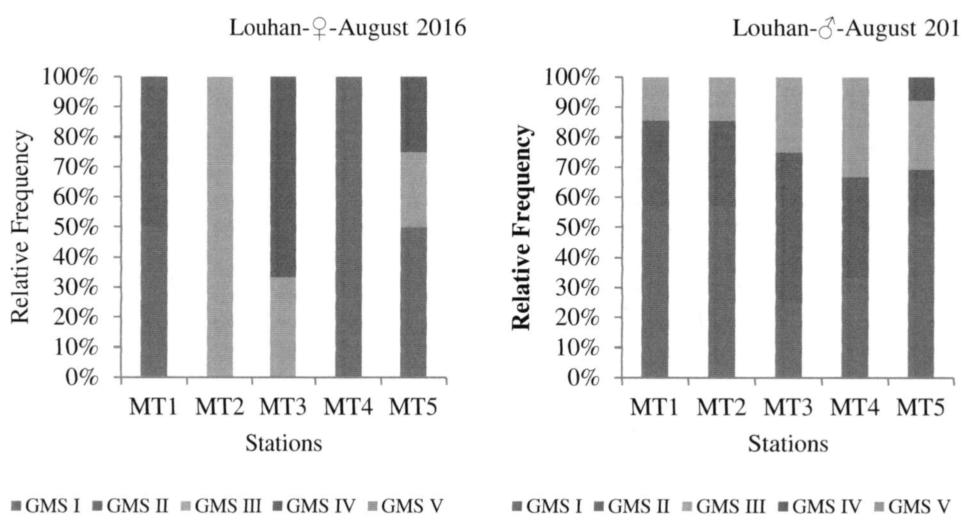

(a) (b)

Figure 15. Gonadal maturity stage of Louhan (a) female and (b) male in August at Lake Matano.

4. Conclusion

Telmatherina antoniae was relatively more abundant in Petea and near Flats, while Louhan was abundant in Nuha and Bubble beach. Female of *T.antoniae* was relatively bigger than male. Length of *T. antoniae* ranged from 69 to 105 mm for female and 64 to 102 mm for male, furthermore the weight ranged from 3.2 to 11 gram for female and 2.3 to 11 gram for male. On the other hand, female of Louhan was relatively smaller than male. Length of female Louhan ranged from 51 to 172 mm and male from 53 to 185 mm. Weight of Louhan ranged from 2.5 to 90.2 gram and 2.5 to 102.6 for female and male, respectively. The growth pattern of overall *T.antoniae*was isometric as well as Louhan.

Both *T. antoniae* and Louhan were found in GMS I to IV with a greater proportion of mature gonads in August.

Alien fish species that tend to be invasive, such as Louhan, must be controlled through mass capture either using a net of at least 2 inches (gonad mature Louhan and opudi fish not caught), fishing rods, or rifles that could be done by the community through government or private programs. So that Louhan fish can be controlled quickly. Moreover, compensate the Louhan catch with some incentive also could be an alternate.

5. References

[1] Wirjoatmodjo S, Sulistiono, Rahardjo M F, Suwelo I S and Hadiyati R K 2003 *Ecological distribution of endemic fish species in Lakes Poso and Malili Complex, Sulawesi Island*. Funded by Asean Regional Centre for Biodiversity Conservation and the European Comissionp 30

[2] Jayadi, Rimal H, Arifuddin 2010 *Torani (Jurnal Ilmu Kelautan dan Perikanan)* **20**(1):44-48

[3] Froese R and Pauly D 2004 FishbaseWorlWide Web electronic publicationwww.fishbase.org, version (06/2004).

[4] Reid W V and Miller K R 1989*Keeping option alive.* The Scientific Basis for Conserving Biodiversity. WRI.

[5] Nasution S H 2006 Pangkilang (Telmatherinidae) ornamental fish: An economic alternative for people surround Lake Towuti. *Proc. Int. Symp. The Ecology and Limnology of the Malili Lakes* Bogor-Indonesia: March 20-22, 2006. Supported by PT. INCO Tbk. and Research Center for Limnology, Indonesian Institute of Sciences (LIPI) pp 39-46

[6] Nasution S H 2007*Indonesian Fisheries Research Journal* **13**(2):117-123

[7] Sparre P and Venema S C 1999 *Introduksi Pengkajian Stok IkanTropis*. Puslitbangkan, penerjemah. Jakarta: Pusat Penelitian dan Pengembangan Perikanan. Terjemahan dari: *Introduction to Tropical Fish Stock Assessment*

[8] King M 1995 *Fisheries Biology, Assessment, and Management* United Kingdom: Fishing News Books

[9] Busacker G P , Adelman I R and Goolish E M 1990 *Growth* pp 363-382 *in* Schreck, C B and Moyle P B (Eds), Methods for Fish Biology Maryland, USA: American Fisheries Society

[10] Effendie M I 1979 *Metoda biologi perikanan* Bogor Yayasan Dewi Sri Cetakan I p 112

[11] Holden M J and Raitt D F S 1974 *Manual of Fisheries Science Part 2: Methods and Reources Investigation and Their Application* Rome: Food and Organization of United Nations.

[12] Weber M and de Beaufort K L F 1913 *The fisheries of Indo-Australia archipelago* Vol. II. Leiden E J Brill. Ltd. p 404

[13] Weber M and de Beaufort K L F 1916 *The fisheries of Indo-Australia archipelago.* Vol. III. Leiden E J Brill. Ltd. p 455

[14] Weber M and de Beaufort K L F 1922 *The fisheries of Indo-Australia archipelago.* Vol. IV. Leiden E J Brill. Ltd. p 410

[15] Kottelat M, Whitten A J, Kartikasari S N and Wirjoatmodjo S 1993*Freshwater fishes of Western Indonesia and Sulawesi.* Jakart: Periplus Edition (HK) Ltd. in collaboration with the Environmental Management Development in Indonesia (EMDI) Project, Ministry of State for Population and Environment, Republic of Indonesia p 293

[16] Pauly D 1984 *Fish Population Dynamics in Tropical Waters:A Manual for Use with Programmable Calculators* Manila: ICLARM

[17] Effendie M I 2002 *Biologi Perikanan* Edisi RevisiYogyakarta:Yayasan Pustaka Nusantara p163

[18] Steel R G D and Torrie J H 1989 *Principles and procedure of statistic* Second Edition New York: Mic Graw Hill Book Company, Inc p 748

[19] Anonim 2001 Peraturan Pemerintah Republik Indonesia Nomor 82 Tahun 2001 tentang pengelolaan kualitas air dan pengendalian pencemaran air

[20] Goldman CR, Horne A J 1983 *Limnology* New York: Mc Graw Hill International Book Company p 464
[21] Tantu F 2012 *Ekobiologi Reproduksi Ikan OpudiTelmatherina antoniae (Kottelat, 1991)sebagai Dasar Konservasi Ikan Endemik di Danau Matano, Sulawesi Selatan* Bogor: Disertasi Sekolah Pascasarjana Institut Pertanian Bogor Bogor p 97
[22] Gray S M and McKinnon J S 2006 *Journal of Biology of Fishes* **75**:471-482
[23] Herder F, Hadiaty R K, Schwarzer J, Pfaender J, Nolte A, Schliewen U2006 Diversity and evolution of Telmatherinidae in the Malili Lakes System (Sulawesi). *Proc. Int. Sym. The Ecology and Limnology of the Malili Lakes*Bogor-Indonesia: March 20-22, 2006 Supported by PT. INCO Tbk. and Research Center for Limnology, Indonesian Institute of Sciences (LIPI). pp 67-72
[24] Sumassetiyadi M A 2003 Beberapa aspek reproduksi ikan opudi (*Telmatherina antoniae*) di Danau Matano Sulawesi Selatan. Program Studi Manajemen Sumberdaya Perairan, Fakultas Perikanan dan Ilmu Kelautan Institut Pertanian Bogor p 55
[25] Dewantoro G and Rachmatika I 2016 *Jenis Ikan Introduksi dan Invasif Asing di Indonesia* Jakarta: LIPI Pressp 192
[26] Merta I G S 1993 *Jurnal Penelitian Perairan Laut* **73** : 35 - 44
[27] Miller P J 1984*The Tokology of Gobioid Fishes*(*in*)Potts G W and Wootton R J (eds.). Fish reproduction, strategies, and tactics London: Academic Press. Harcourt Brace Jovanovich Publishers pp 223-244.
[28] Nasution S H 2005 *Jurnal Penelitian Perikanan Indonesia. Edisi Sumber Daya dan Penangkapan* **11** (2):29-37
[29] Nasution S H, Said D S, Lukman, Triyantoand Fauzi H 2006 Aspek reproduksi ikan beseng-beseng (*Telmatherina ladigesi*Ahl) dari beberapa sungai di Sulawesi Selatan. Prosiding Seminar Nasional Ikan IV Jatiluhur:29-30 Agustus 2006 pp 83-93
[30] Nagahama Y 1983 The Functional Morphology of Teleost Gonads. *In* Hoar W S, Randal D J and Donaldson E M (eds.). *Fish physiology*, IX A: 223-276. Academic Press, New York.
[31] Hardjamulia A, Suhenda N and Wahyudi E 1995 *Jurnal Penelitian Perikanan Indonesia* **1**(3):36-46
[32] Takashima F and Hibiya T 1995*An Atlas of Fish Histology Normal and Phatological Features.* 2 nd edition Tokyop 195
[33] Yoneda M, Futagawa K, Tokimura M, Horikawa H, Matsuura S and Matsuyama M 2002 *Fisheries Research* **57**:297-309

Acknowledgments
This study was a part of research project titled "*Adaptasi dan Mitigasi Perubahan Iklim pada Perairan Darat. Studi Kasus: Ekosistem Perairan Danau dan Situ*" (Adaptation and Mitigation of Climate Change in Inland Waters. Case Study: Lake Ecosystem) funded by *Unggulan LIPI* programme. We thank to all team members who helped this research.

Effects of asymmetrical biotic interactions on multispecies dynamics

J O Ojonubah[1*] and M H Mohd[1]

[1] School of Mathematical Sciences, Universiti Sains Malaysia, 11800, USM, Penang, Malaysia

[*] *Corresponding author: ojonubahjames@gmail.com*

Abstract. One of the key challenges in ecology is to predict multispecies spatial distributions in the future. The influence of biotic interactions in the form of competition and environmental change such as climate are thought to have significant impacts on the distribution of species. To ascertain this, we study a multispecies community assembly with asymmetrical competition across heterogeneous environments using mathematical modelling. With the aid of analytical and simulation analyses, we achieved the predictions of the species range margins (comprising the geographic boundary beyond which a species is not present). We also perform bifurcation analysis in order to understand the competitive dynamics across location. Our results show that the combined influence of the biotic and abiotic factors lead to coexistence of species when competition is relatively weak and existence of single dominant species territory during aggressive competitions. Also, the threshold values observed in the bifurcation analysis correspond to transcritical bifurcation points.

1. Introduction

In nature, species range margins and biodiversity across geographical locations may be changing due to the combine influences of ecological factors such as climate change and biotic interactions. The consequence is that the prediction of the future presence-absence of species in a geographical location may be a challenging task to ecologists. This aspect is thought to have dramatic effects on the conservation plans that can alleviate biodiversity. Therefore, anticipation and management of these effects are among the most crucial problems in ecology [1, 2]. As such, the presence-absence of species by which we refer to the boundary where a species is present or absent and also means the range margins or limits of species is one of the main thrusts of this study. In this study, we are also concerned with ecological forces that influence species community assemblies across geographical region and the mechanisms that can induce coexistence of species and the occurrence of priority effects. Environmental factors (such as climate change) and biotic interactions (such as interactions among species) [2, 3] are known to have significant impacts on the distributions and range margins of species. Although range shifts of species have earlier been attributed to biotic interactions by researchers in the past [4], emphases are frequently placed on climate change by contemporary ecologists to be another responsible factor [5-7]. Generally, species over time and space, respond differently to environmental changes and as well as biotic interactions [8]. Therefore, since biotic interactions among species simultaneously take place with the changes in the environment, species distribution study which focus on the joint effects of climate change and biotic interactions, may be in the right direction.

IOP Conf. Series: Earth and Environmental Science **380** (2019) 012013 doi:10.1088/1755-1315/380/1/012013

There are several empirical evidences which inspired this study to show that species distribution can jointly be shaped by biotic interactions and climate change. For instance, Connell [9] investigated interspecific competition between Balanus and *Chthamalus* species in rocky intertidal zone along Scottish coast. He observed vertical distribution between the two species due to competition where *Chthamalus* occupied the upper region but was limited from above by desiccation and Balanus occupied the immediate region below. In addition, Park *et al.* [10] conducted experimental studies which also show the importance of interaction between biotic factor and abiotic environments (e.g. temperature and humidity) can determine species distribution with the occurrence of priority effects. According to Gilman *et al.* [11] climate change effect on species can intensely be determined by species biotic interactions. Also, Meier *et al.* [12] stated that species interactions with each other can impact species response to changes in the environment and in a similar way, environmental changes can as well determine the trend of species interactions in a community. These observations may be due to the fact that factors that affect species interactions either directly or indirectly most often vary systematically across environmental gradients [13].

However, how biotic interactions combined with environmental factors to influence species distributions in their community remains not clear and need to be investigated further. For instance, Wisz *et al.* [2] reviewed examples where plant distributions are shaped by biotic interactions, agreed that we lack the data to estimate the relative significance of biotic and abiotic factors in plants. Nevertheless, the empirical limitations can be resolved through mathematical modelling approach. There are existing mathematical models which give better understanding of how biotic interactions can influence distributions of species [14-17]. In addition, Godsoe *et al.* [13] and Mohd *et al.* [18] investigated the combined effects of biotic interactions and abiotic environments on the range margins of species using modelling approach. However, few of these studies consider the asymmetrical interactions aspect of species competitions. For instance, the range margin predictions from the model assumptions of [13, 18] that interspecific competition is symmetric may require further investigations. They assumed equal competition strengths for all the species which is rarely true in nature. According to Roughgarden [19] asymmetrical models prediction differ from those of symmetrical models. In this regards, Mohd *et al.* [18] further suggested that the analysis of the range margins of species can be carried out with consideration for asymmetric competitions. The empirical research on snail species in two streams in the western United States to measure biotic interaction strengths between the snail species by [20] demonstrates that asymmetric interactions could be responsible for the patterns of dominance by one of the snails species. Also, Morin [21] reported that asymmetric strengths of each type of competitions can lead to different outcomes.

In this view, asymmetrical competition models may give competition outcomes different from that of symmetrical models. The question is, does asymmetrical competition model really give different predictions from symmetrical models? To gain insights into this question, we extend previous mathematical models of [13] and [18], which model biotic interactions and climate change with the assumption that biotic interactions among species are symmetrical. The purpose is to investigate additional competition outcomes that may arise from an asymmetrical model. We focus on competition between multiple species, example, four species. Thus, our model consists of four systems of ordinary differential equations, which represent asymmetric interspecific competitions among four species in a heterogeneous environment. To model abiotic components, we incorporate the suitability of the environments into the carrying capacities of each species. The inclusion therefore combined the effects of biotic interactions and abiotic factors (environmental gradients) on multispecies community dynamics.

This study is organised as follows. After this introduction is the model descriptions which combine biotic and abiotic factors. This is followed by analytical results on the range margins of species, simulation results and the bifurcation analysis. Based on our results, we discuss the ecological implications of the results.

2. Methodology

2.1. Species Competition Model

To predict the distribution of competing species across heterogeneous environments, we use mathematical model for the densities $N_i(x,t)$ of *n* species. The model is an extension of Lotka-Voltera competition model [22, 23] extended along one dimensional environmental gradient *x*, where $0 \le x \le 1$ such that:

$$\frac{dN_i}{dt} = \frac{r_i N_i}{k_i(x)}(k_i(x) - \sum_{j=1}^{n} \alpha_{ij} N_j); \qquad (i = 1,2,\ldots,n) \tag{1}$$

where r_i is the intrinsic growth rate, $k_i(x)$ is the carrying capacity, α_{ij} is the strength of competition of species *j* on species *i*, α_{ii} is the intraspecific coefficients and N_i is the densities of species *i* at time *t*. By rescaling the density of species *i* relative to its competition coefficients α_{ii}, we set α_{ii} to equal 1, such that competition coefficients, α_{ij} signify the ratio of interspecific to intraspecific competition.

The effect of environmental gradients on species competition outcomes is determined by $k_i(x)$ in the presence of biotic interactions. Thus, the maximum number of species that an environment can accommodate at a time, depends on the carrying capacity of the environment. In this case, *x* is the varying parameter of $k_i(x)$ and it can represent the abiotic environmental features like climate, humidity, salinity and any other environmental factors that affect the species in a geographical location. In this study, our model is a reflection of laboratory research of [24, 25] such that *x* strictly denote temperature. Therefore, the effects of biotic interactions on range margins of species may depend on how the species respond to environmental factors. To show these effects in the interactions of multispecies communities, the carrying capacity term (i.e $k_i(x)$) is modelled such that it varies linearly with *x* as shown in (2):

$$k_i(x) = m_i x + b_i \tag{2}$$

Here c_i is the intercept of species *i* carrying capacity when $x = 0$, m_i is the slope of species *i* carrying capacity which is a measure of a change in the suitability of the environment with respect to abiotic component *x* and $k_i(x)$ is as earlier defined. Since *x* is a function of the carrying capacity $k_i(x)$ of species *i*, it's therefore restricted by the abiotic environment; such that the distribution of the species *i* in the absence of competition reaches its maximum at $x = 1$ for all positive m_i. Thus, for four competing species (i.e $n = 4$), equation (1) becomes:

$$\begin{aligned}
\frac{dN_1}{dt} &= \frac{r_1 N_1}{k_1}(k_1 - N_1 - \alpha_{12}N_2 - \alpha_{13}N_3 - \alpha_{14}N_4)\\
\frac{dN_2}{dt} &= \frac{r_2 N_2}{k_2}(k_2 - N_2 - \alpha_{21}N_1 - \alpha_{23}N_3 - \alpha_{24}N_4)\\
\frac{dN_3}{dt} &= \frac{r_3 N_3}{k_3}(k_3 - N_3 - \alpha_{31}N_1 - \alpha_{32}N_2 - \alpha_{34}N_4)\\
\frac{dN_4}{dt} &= \frac{r_4 N_4}{k_4}(k_4 - N_4 - \alpha_{41}N_1 - \alpha_{42}N_2 - \alpha_{43}N_3)
\end{aligned} \tag{3}$$

In equation (3), the model represents the population dynamics of competing species such that N_1, N_2, N_3 and N_4 respectively represent species 1- 4 densities. For model simplification, we will consider $\alpha_{ij} = \alpha_j$ in our analysis and so, α_1, α_2, α_3 and α_4 denote species 1 - 4 competition coefficients.

2.2 Steady states and stability analysis

To understand the dynamics of the system, we compute the steady states of equation (3) and then analyse the stability of the steady states. The steady states are computed by setting $\frac{dN_1}{dt}$, $\frac{dN_2}{dt}$, $\frac{dN_3}{dt}$ and $\frac{dN_4}{dt}$ to zeros and the stability of the steady states are performed with the aid of MAPLE package. Thus, only the relevant stable steady states will be discussed here. The stability status is obtained by analysing the nature of eigenvalues of the Jacobian matrices and since $k_i(x)$ are functions of *x*, the stability analysis of

IOP Conf. Series: Earth and Environmental Science **380** (2019) 012013 doi:10.1088/1755-1315/380/1/012013

the steady states depends on the location x. Therefore, a steady state with all negative real parts of the eigenvalues is stable.

2.3 Numerical methods

Based on the steady states, we used the techniques of invasion analysis to derive analytical results on species' range margins for model (3). Numerical simulation results on the range margins of the species are obtained by employing MATLAB ode15 solver for $t = 1000$ to solve model (3); i.e until steady states were achieved. The stability of the steady states is checked using MAPLE package. The numerical simulations were carried out separately for relatively weak ($\alpha_{ij} < 1$) and relatively strong ($\alpha_{ij} > 1$) interspecific competition cases. The parameters values used for our analyses are presented in Table 1. The numerical continuation package XPPAUT to compute bifurcation diagrams of model (3) and the stable and unstable steady states are tracked using AUTO as competition coefficient (α_1) changes. The bifurcation analyses illustrate occurrence of different threshold phenomena, which determine the outcomes of biotic interactions such as species coexistence and priority effects.

Table 1. Symbols with the descriptions and parameter values used for computation of figures.

Symbol	Items description	Parameter value
r_i	Intrinsic growth rates of species i	1
m_1	Gradient of k_1	1
m_2	Gradient of k_2	0.8
m_3	Gradient of k_3	-1
m_4	Gradient of k_4	-0.8
b_1	Carrying capacity of species 1 at $x = 0$	0
b_2	Carrying capacity of species 2 at $x = 0$	0
b_3	Carrying capacity of species 3 at $x = 0$	1
b_4	Carrying capacity of species 4 at $x = 0$	0.8

3. Results

3.1 Analytical results on Range margins of species

In this section, we derive and present analytical results of the model (3) using the method of invasion analysis. The results are specifically based on illustration from figure 1A. The method of invasion analysis is such that a species can invade if the growth rate of the species is greater than zero (i.e $\frac{dN_i}{dt} > 0$) [26]. Thus, boundary invasion of an invading species requires that $\frac{dN_i}{dt}$ in (3) to be greater than zero for the species to invade. For instance, at the invasion point of species 1, only species 3 and 4 are present and so, species 1 can only invade if

$$k_1(x) > \alpha_3 N_3^* + \alpha_4 N_4^* \quad (4)$$

and the point x where $k_1(x) = \alpha_3 N_3^* + \alpha_4 N_4^*$ satisfy the invasion point of species 1; which corresponds to N_3^* *and* N_4^* stabilities. Since species 1 and 2 are absent at the invasion point of species 1, their densities (i.e N_1 and N_2) are taken to be zero at the invasion point of species 1. In this case, N_1 and N_2 are unstable at the species 1 invasion point. From our stability analysis results, the steady state

$(0,0,N_3^*,N_4^*)=(0,0,\frac{\alpha_4 k_4-k_3}{\alpha_3\alpha_4-1},\frac{\alpha_3 k_3-k_4}{\alpha_3\alpha_4-1})$ where $k_3=m_3x+b_3$, $k_4=m_4x+b_4$ and $k_1=m_1x$; when substituted into equation (4) gives the invasion point of species 1 as

$$x_1=\frac{(\alpha_3\alpha_4-\alpha_3)b_3+(\alpha_3\alpha_4-\alpha_4)b_4}{(\alpha_3\alpha_4-1)m_1-(\alpha_3\alpha_4-\alpha_3)m_3-(\alpha_3\alpha_4-\alpha_4)m_4} \tag{5}$$

In a similar way, species 2 invasion point corresponds to the presence of species 1, 3 and 4 in the simulation results such that species 2 can invade if

$$k_2(x)>(\alpha_1N_1^*+\alpha_3N_3^*+\alpha_4N_4^*) \tag{6}$$

and the point where $k_2(x)=(\alpha_1N_1^*+\alpha_3N_3^*+\alpha_4N_4^*)$ satisfies the invasion point of species 2. In this case, the only stable steady state becomes

$$(N_1^*,0,N_3^*,N_4^*)=(\frac{-\alpha_3\alpha_4k_1+\alpha_3\alpha_4k_3+\alpha_3\alpha_4k_4-\alpha_3k_3-\alpha_4k_4+k_1}{2\alpha_1\alpha_3\alpha_4-\alpha_1\alpha_3-\alpha_1\alpha_4-\alpha_3\alpha_4+1},0,\frac{\alpha_1\alpha_4k_1-\alpha_1\alpha_4k_3+\alpha_1\alpha_4k_4-\alpha_1k_1-\alpha_4k_4+k_3}{2\alpha_1\alpha_3\alpha_4-\alpha_1\alpha_3-\alpha_1\alpha_4-\alpha_3\alpha_4+1},\frac{\alpha_1\alpha_3k_1+\alpha_1\alpha_3k_3-\alpha_1\alpha_3k_4-\alpha_1k_1-\alpha_3k_3+k_4}{2\alpha_1\alpha_3\alpha_4-\alpha_1\alpha_3-\alpha_1\alpha_4-\alpha_3\alpha_4+1}),$$

with $k_3=m_3x+b_3$, $k_4=m_4x+b_4$ and $k_2=m_2x$. Thus, species 2 invasion point on substitution is given as

$$x_2=\frac{(\alpha_1\alpha_3\alpha_4-\alpha_1\alpha_3-\alpha_3\alpha_4+\alpha_3)b_3+(\alpha_1\alpha_3\alpha_4-\alpha_3\alpha_4-\alpha_1\alpha_4+\alpha_4)b_4}{(2\alpha_1\alpha_3\alpha_4-\alpha_1\alpha_3-\alpha_1\alpha_4-\alpha_3\alpha_4+1)m_2-(\alpha_1\alpha_3\alpha_4-\alpha_1\alpha_3-\alpha_1\alpha_4+\alpha_1)m_1-(\alpha_1\alpha_3\alpha_4-\alpha_1\alpha_3-\alpha_3\alpha_4+\alpha_3)m_3-(\alpha_1\alpha_3\alpha_4-\alpha_3\alpha_4-\alpha_1\alpha_4+\alpha_4)m_4} \tag{7}$$

As we move towards the right along the environmental gradients x are the invasion points of species 4 and then species 3 which respectively correspond to the presence of species 2, 3 and 4 and then species 1 and 2. Thus, we can derive the invasion point of species 3 and 4 similar to that of species 1 and 2 to give

$$x_4=\frac{(2\alpha_1\alpha_2\alpha_3-\alpha_1\alpha_2-\alpha_1\alpha_3-\alpha_2\alpha_3+1)b_4-(\alpha_1\alpha_2-\alpha_1-\alpha_2+1)\alpha_3b_3}{(\alpha_2\alpha_3-\alpha_2-\alpha_3+1)\alpha_1m_1+(\alpha_1\alpha_3-\alpha_1-\alpha_3+1)\alpha_2m_2} \tag{8}$$

and

$$x_3=\frac{(\alpha_1\alpha_2-1)b_3}{(\alpha_2-1)\alpha_1m_1+(\alpha_1-1)\alpha_2m_2} \tag{9}$$

All the invasion points stated above correspond to transcritical bifurcation points where each species exchange its stability from being unstable to become stable.

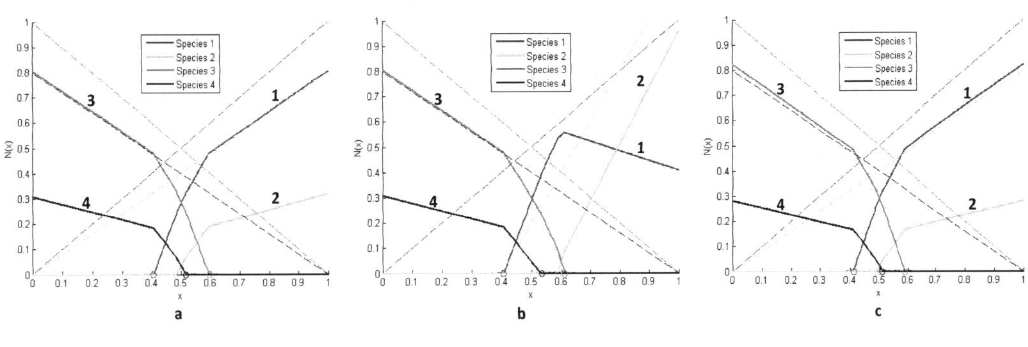

Figure 1. The steady states of equation (3) resulting from environmental gradients (2). In both figures, solid lines indicate steady states of species i and the dotted lines represent their carrying capacities (i.e $k_i(x)$). The invasion points (x_i) of the species are on the horizontal axis. Figures 1a and b represent asymmetric competition with $\alpha_1=0.6$, $\alpha_2=0.62$, $\alpha_3=0.61$, $\alpha_4=0.63$ and figure 1c represents symmetric competition with $\alpha_i=0.63$. Initial abundance used for the figures: $N_1(x)=0.1k_1(x)$, $N_2(x)=0.9k_2(x)$, $N_3(x)=0.1k_3(x)$, $N_4(x)=0.9k_4(x)$. Other parameters values used for the computation of these results are

given in table 1 except in figure 1b where species 2 carrying capacity is $k_2(x) = 2x - 0.8$ and MATLAB ode 15s solver is used to compute the results.

3.2 Range margins of species for relatively weak competitions (i.e $\alpha_i < 1$)

Figure 1a and b shows the asymmetrical competition outcomes of four species for $\alpha_i < 1$ as predicted by equation (3) and figure 1c is presented to illustrate symmetrical competition outcomes for comparison with asymmetrical results. Parameter values used to illustrate the dynamical outcomes of the competitions are listed in table 1 and in figure 1 caption. In figure 1a, the range margins of species 1 for example is at $x = 0.4068$ (blue circle); compared with figure 1c where the range margins of the species 1 is at $x = 0.412$ (blue circle). The range margins of the four species are on the x-axis for species 1 – 4; respectively labelled as blue, green, black and red circles. The numerical results of the range margins of the four species computed for only asymmetrical competition, correspond to the analytical results of the range margins (x_i) stated in (5), (7), (8) and (9). As a result, the spatial domain of the species is divided into regions of two species, three species and four species coexistence. In figure 1a for instance, the regions of two species coexistence correspond to two domains. Thus, from the left, we have the region (i.e $x \leq x_1$) where species 3 and 4 coexist and displace species 1 and 2 to invasion points of 0.4068 and 0.4845 respectively. Also, from the right is the region (i.e $x_3 \leq x$) where species 1 and 2 coexist and displace species 3 and 4 to extinction point of 0.5956 and 0.5161 respectively. These are followed by two regions (i.e $x_1 \leq x \leq x_2$ and $x_4 \leq x \leq x_3$) of three species coexistence. In the first region, species 1, 3 and 4 are present and exclude only species 2. Similarly, in the second region, species 1, 2 and 3 are present and exclude only species 4. The last region is the middle domain (i.e $x_2 \leq x \leq x_4$), where all the four species coexist. These observations illustrate that for relatively weak competition, coexistence of two (2) or more species is possible. In this case, both the carrying capacity and competition strengths of the species combined to determine the dynamical competition outcomes of the species.

Similar observations are possible in the case of figure 1b, which is computed to illustrate the importance of environmental gradients in the species competition dynamics. In order to demonstrate this significance, only species 2 environmental gradient is altered from $k_2 = 0.8x$ to $2x - 0.8$; while other parameters remain the same with figure 1a. In this way, the community assemblies observed in figure 1a are not possible in figure 1b due slight change in the environmental gradient of species 2. As a result, the order of invasion points (x_i) of the species in figure 1b becomes x_1, x_4, x_2, and x_3; respectively labelled as blue, black, green and red circles. In this case, multispecies (i.e four species) coexistence is not possible which is quite different from our observations in figure 1a.

Also, figure 1c shows a similar qualitative result when compared with figure 1a. However, their range margins predictions differ. For example, species 1 invasion point in figure 1a is at $x = 0.4068$ which is shifted to $x = 0.412$ in figure 1c. As a result, species 3 is observed to occupy a region beyond species 4 carrying capacity (compare figure 1a and 1c).

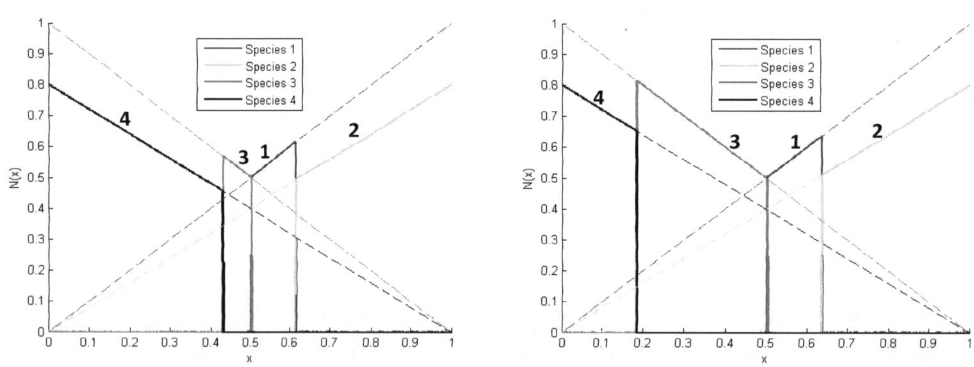

Figure 2. The steady state of model (3) following environmental gradient (2). Solid lines indicate steady states of species i and the dotted lines indicate their carrying capacities. Figure 2a and b represent relatively strong competitions with: $\alpha_1 = 1.10$ $\alpha_2 = 1.31$ $\alpha_3 = 1.11$, $\alpha_4 = 1.32$. Figure 2a initial abundance: $N_1(x) = 0.1k_1(x)$, $N_2(x) = 0.9k_2(x)$, $N_3(x) = 0.1k_3(x)$, $N_4(x) = 0.9k_2(x)$ and 2b initial abundance: $N_1(x) = 0.08k_1(x)$, $N_2(x) = 0.7k_2(x)$, $N_3(x) = 0.1k_3(x)$, $N_4(x) = 0.7k_2(x)$.

3.3 Range margins of species for relatively strong competitions (i.e $\alpha_i > 1$)

Figure 2a and b illustrate the model (3) predictions on four species competition outcomes for relatively strong competitions (i.e $\alpha_i > 1$). Parameter values used to illustrate the dynamical outcomes of the competitions are outlined in table 1 and in figure 2 caption. For this case of strong competition, coexistence of species is impossible such that the dynamical behaviour of the model leads to exclusion of other competitors from a particular region with only dominant species occupying the region to it carrying capacity. In this situation, the outcomes of the competition depend on competition strength and initial abundance of the species [27]; where initially more abundant species have more potentials to dominate the competition.

Thus, figure 2a and b are respectively computed to illustrate the impacts of competition strength and initial abundance on competition outcomes. As a result of aggressive competitions among the four species, the spatial domain in the two figures are divided into four regions of single species each to it carrying capacity. Observe that in figure 1a, the regions of single species from the left correspond to the domain (black) where only species 4 is present and dominates species 1, 2 and 3. This is followed by the region (red) with only species 3 present. Also, to the right of this region is the species 1 dominant domain (blue) where it excludes other competitors. The last region on the right, is where species 2 dominates (green) the competition and exist as single species. Similarly, in figure 2b the spatial domain is also divided into regions of single species, but the width of the area occupied by each species either increases or decreases as initial abundance of the species is varied compared with figure 2a.

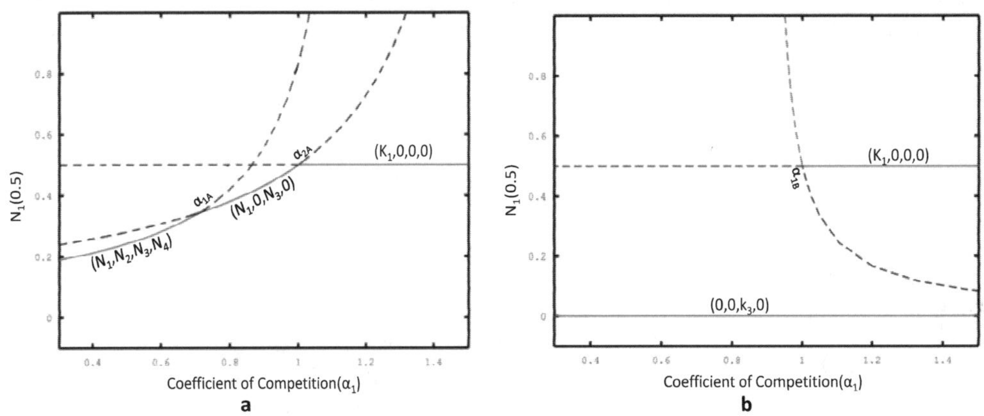

Figure 3. The bifurcation analysis of density steady state of species at a location $x = 0.5$ for equation (3) resulting from environmental gradients (2) with varied species 1 competition strength α_1. Solid red lines represent stable steady states, dotted red line denote infeasible solution (i.e one of the densities of species is negative) and dotted black lines indicate unstable steady states. Figure 3a is computed using: $\alpha_1 = 0.3\text{-}1.5$, $\alpha_2 = 0.62$, $\alpha_3 = 0.61$, $\alpha_4 = 0.63$; while figure 3b is computed using: $\alpha_1 = 0.3\text{-}1.5$, $\alpha_2 = 1.12$, $\alpha_3 = 1.11$, $\alpha_4 = 1.13$ and other parameters values used for the computation of the results are given in table 1. The threshold values correspond to transcritical bifurcation points (i.e α_{1A}, α_{2A} and α_{1B}) and XPPAUT is used to compute the results with the aid of AUTO.

3.4 Bifurcation analysis

We performed bifurcation analyses at a location $x = 0.5$ in order to further understand the diverse species that are present or absent at this location x as species 1 competition strength α_1 changes. The bifurcation analysis, therefore, reveals the threshold value of species 1 competition strength α_1. Thus, figure 3a and b is constructed to track the stable and unstable steady states of the species that are present or absent in model (3). Figure 3a is the model bifurcation outcomes of varying species 1 competition strength α_1 while other species competition strengths are kept constant at relatively weak competitions (i.e below α_i

< 1). Similarly, in figure 3b the competition strengths of other species are kept constant at relatively strong competitions (i.e above $\alpha_i > 1$) while species 1 competition strength α_1 can vary. In both figures the solid red lines represent the stable steady states densities of species 1 relative to other competitors, dotted red line denote infeasible solution (i.e one of the densities of species is negative) and the dotted black lines denote unstable steady states of species.

We observe that figure 3a and figure 3b give different dynamics at the location $x = 0.5$. Thus, we have in figure 3a coexistence of the four species at relatively weak competition α_1 and exclusion of species as competition gets stronger; which lead to coexistence of only two species (i.e species 1 and 3) and then single-species (i.e species 1). In the case of figure 3b, we have existence of single-species (i.e species 3) and as competition becomes stronger, priority effects occur where two different single-species (i.e species 3 and species 1) exist at the same location x. Notice in figure 3a that coexistence of species is only possible at $\alpha_i < 1$ and as competition strength becomes aggressive (i.e $\alpha_i > 1$) all other competitors are excluded with only species 1 existing as single species to its' carrying capacity. However, due to impacts of strong competitions from other species in figure 3b, coexistence of species is impossible at the location. Also, the threshold values (i.e α_{1A} α_{2A} *and* α_{1B}) in the figures corresponds to transcritical bifurcation points where one combination of species exchange its stability for another combination of species.

4. Discussion

We study the combined effects of biotic interactions and environmental factors on multispecies asymmetric competition outcomes in heterogeneous environments. Although, it is no longer a new idea that environmental factors such as climate change [5-7] and biotic interactions [2, 3, 28] can individually impacts species range margins, but their studies is still relevant in ecology. This is because, the extent to which the two factors relatively combined to determine species distribution (i.e the presence-absence) in a geographical location is persistently a challenge in ecology [2]. However, through mathematical modelling approach, we show that multispecies community assembly can be influenced by combined influence of biotic interactions and environmental factors. Our finding is robust because, it shows that both the two factors are significant in species distribution. As a result, our model prediction revealed coexistence of species at the central region of the geographical location than at the peripheral of the location where exclusion of species is observed even at relatively weak competition. This form of species distribution was also observed in the empirical study of small mammal species along elevational gradients. This is because, some species that are not favoured in an environment can easily be excluded in the presence of biotic interactions; such that multispecies coexistence is only possible where all the species are favoured by the environment.

When competition becomes aggressive, we observed that competition outcomes are determined by the initial abundance of the species and can be influenced by species competition strengths. This observation may add value to existing knowledge which claimed that initial abundance of the species is the sole determinant of competition outcomes during intense competition [13, 18]. It is possible that species competition strength to be the umpire of competition alongside with initial abundance depending on parametrization. In this case, only one species dominates the competition in a particular location and exist as a single species to its' carrying capacity whose area occupied can be determine by both initial abundance and competition strength.

These observations are qualitatively consistent with symmetrical biotic interactions of multispecies dynamics [18]. However, the range margins predictions of asymmetrical biotic interactions of multispecies differ quantitatively from that of symmetrical interactions. Such variation is attributed to unequal interaction strengths with which the species interact with one another, which obviously, is not the case in symmetrical interactions.

Our work also revealed the importance of tracking the stable and unstable steady states of the model using bifurcation analysis. This is because, bifurcation analysis gives detailed theoretical explanation of the observed differences of species presence-absence in our numerical simulation results at a particular location x, while competition strength changes. Our bifurcation results revealed existence of threshold values of competition strengths which lead to coexistence and priority effects. The threshold values correspond to transcritical bifurcations which are the points where one combination of species gradually exchange its' stability for another one.

5. Conclusion

Environmental change may continue to impact on range margins of species community assembly. As a result, identifying that biotic interactions can mediate species responses to environmental effects is important in multispecies ecological dynamics. This has been achieved through our theoretical models which is used to study the combined effects of biotic interactions with the environmental factors. Consequently, our result shows that the combined influence of biotic and abiotic factors can lead to coexistence of species when competition strengths are relatively weak. However, in the presence of relatively aggressive competition, mediated by the initial abundance of the species, coexistence is impossible; such that only dominant single species occupying a particular region to it carrying capacity. The challenge for researchers in this direction, is to extend this model for accurate predictions and forecasts of future species distributions under climate change. Efforts in this direction, requires appropriate model assumptions and parameterization.

6. References

[1] Sexton J P, McIntyre P J, Angert A L, and Rice K J 2009 Evolution and ecology of species range limits *Annual Review of Ecology, Evolution, and Systematics* **40** pp 415–36

[2] Wisz M S, Pottier J, Kissling W D, Pellissier L, Lenoir J, Damgaard C F, Dormann C F, Forchhammer M C, Grytnes J A and Guisan A 2013 The role of biotic interactions in shaping distributions and realised assemblages of species: implications for species distribution modelling *Biological Reviews* **88**(1) pp 15-30

[3] Godsoe W, Murray R and Plank M J 2015 The effect of competition on species' distributions depends on coexistence, rather than scale alone *Ecography* **38**(11) pp 1071-79

[4] Grinnell J 1917 The niche-relationships of the California Thrasher *The Auk* **34**(4) pp 427-33

[5] Shmida A and Wilson M V 1985 Biological determinants of species diversity *Journal of biogeography* pp 1-20

[6] Pearson R G and Dawson T P 2003 Predicting the impacts of climate change on the distribution of species: are bioclimate envelope models useful? *Global ecology and biogeography* **12**(5) pp 361-71

[7] Soberón J 2007 Grinnellian and Eltonian niches and geographic distributions of species *Ecology letters* **10**(12) pp 1115-23

[8] Algar A C, Kharouba H M, Young E R and Kerr J T 2009 Predicting the future of species diversity: macroecological theory, climate change, and direct tests of alternative forecasting methods *Ecography* **32**(1) pp 22-33

[9] Connell J H 1961 The influence of interspecific competition and other factors on the distribution of the barnacle *Chthamalus* stellatus *Ecology* **42**(4) pp 710-23

[10] Park T, Mertz D B, Grodzinski W and Prus T 1965 Cannibalistic Predation in Populations of Flour Beetles *Physiological Zoology* **38**(3) pp 289-321

[11] Gilman S E, Urban M C, Tewksbury J, Gilchrist G W and Holt R D 2010 A framework for community interactions under climate change *Trends Ecol. Evol.* **25**(6) pp 325-31

[12] Meier E S, Edward T C, Kienast F, Dobbertin M and Zimmermann N E 2011 Co-occurrence patterns of trees along macro-climatic gradients and their potential influence on the present and future distribution of Fagus sylvatica L *Journal of Biogeography* **38**(2) pp 371-82

[13] Godsoe W, Murray R and Plank M J 2014 Information on biotic interactions improves transferability of distribution models *The American Naturalist* **185**(2) pp 281-90

[14] Roughgarden J 1979 *Theory of population genetics and evolutionary ecology: an introduction* Macmillan Publishing

[15] Case T 2000 *An illustrated guide to theoretical ecology. Predator–prey systems: predator dynamics and effects on prey* (New York: Oxford University Press)

[16] Case T J, Holt R D, McPeek M A and Keitt T H 2005 The community context of species' borders: ecological and evolutionary perspectives *Oikos* **108**(1) pp 28-46

[17] Holt R D and Barfield M 2009 Trophic interactions and range limits: the diverse roles of predation *Proc. of the Royal Society of London B- Biological Sciences* vol 276(1661) (London: Biological Sciences) pp 1435-42

[18] Mohd M H, Murray R, Plank M J and Godsoe W 2017 Effects of biotic interactions and dispersal on the presence-absence of multiple species *Chaos, Solitons & Fractals* **99** pp 185-194

[19] Roughgarden J 1972 Evolution of niche width *The American Naturalist* **106**(952) pp 683-718

[20] Riley L A, Dybdahl M F and Hall R O 2008 Invasive species impact: asymmetric interactions between invasive and endemic freshwater snails *Journal of the North American Benthological Society* **27**(3) pp 509-20

[21] Morin P J 2009 *Community ecology* (John Wiley & Sons)

[22] Lotka A J 1924 *Elements of Physical Biology, reprinted 1956 as Elements of Mathematical Biology* (New York: Dover Publishers)

[23] Volterra V 1926 Fluctuations in the abundance of a species considered mathematically *Nature* **118** pp 558-60

[24] Davis A J, Jenkinson L S, Lawton J H, Shorroks B and Wood S 1998 Making mistakes when predicting shifts in species range in response to global warming *Nature* **391**(6669) p 783

[25] Davis A J, Jenkinson L S, Lawton J H and Shorroks B 1998 Individualistic species responses invalidate simple physiological models of community dynamics under global environmental change *Journal of Animal Ecology* **67**(4) pp 600-12

[26] Hastings A and Gross L 2012 *Encyclopedia of theoretical ecology* (Univ of California Press)

[27] Shurin J B, Amarasekare P, Chase J M, Holt R D, Hoopes M F and Leibold M A 2004 Alternative stable states and regional community structure *Journal of Theoretical Biology* **227**(3): pp 359-68

[28] Gilman S E, et al., 2010 A framework for community interactions under climate change *Trends in ecology & evolution* **25**(6) pp 325-31

Acknowledgements

Thanks to the School of Mathematical Sciences and the Universiti Sains Malaysia for the support. Mohd Hafiz Mohd is supported by the Universiti Sains Malaysia Short-Term Research Grant (Grant No. 304/PMATHS/6315083).

Conservation of Agricultural Soil Using Entomopathogenic Fungi: An Agent With Insecticides Degradation Potential

S Q Ong[1*], H Ahmad[2], A H Ab Majid[2] and Z Jaal[2]

[1]KDU Penang University College 32, Jalan Anson, 10400 Penang Malaysia
[2]School of Biological Sciences, Universiti Sains Malaysia, 11800 USM, Penang, Malaysia

**Corresponding author: songguan26@gmail.com*

Abstract. A major current focus in agricultural soil conservation is to ensure a pest control program is sustainable, and therefore, entomopathogenic fungi have been considered and extensively studied as biopesticides. However, the ecological role of entomopathogenic fungi in degrading insecticides in soil is not well understood. In this study, the potential of entomopathogenic fungi *Metarhizium anisopliae* (Met.) in degrading two common agricultural insecticides, chlorpyrifos and cypermethrin was investigated by introducing *M. anisopliae* into autoclaved soils artificially contaminated with 500 ppm of chlorpyrifos, and cypermethrin. The concentration of chlorpyrifos and cypermethrin were determined after 21 days using UPLC/PDA detector. The residues, rate, and percentage of degradation between insecticides treated and control soil were compared using an independent t-test (SPSS 20.0). The degradation of both insecticides in Met. treated soil (>80%) was significantly higher than control soil (47-61%). The residues for chlorpyrifos and cypermethrin residue in Met. treated soils were 19.39±0.10 ppm and 19.68±0.36 ppm, respectively, significantly lower than control (residues of chlorpyrifos-262.6±7.6 ppm and cypermethrin-194.4±4.3 ppm, at $p<0.05$). The results suggested *M. anisopliae* may play a role in the bioremediation of soil.

1. Introduction

Insecticides contribute significantly to the success of modern farming and food production. Although biological control approaches have been commercially applied for some insect species [1], we still rely heavily on insecticides to limit insect damage in agriculture [2]. Insecticides such as chlorpyrifos and cypermethrin were commonly used in crops like corn, pepper, and potatoes, due to its capable effect in controlling soil-dwelling pest. Organophosphorus pesticides such as chlorpyrifos were once regarded as the most widely used insecticides and occupied an estimated 34% of worldwide insecticide sales [3]. However, the insecticides were highly persistent in the ecosystem. This therefore introduces possible incorporation into our food chain, thus affecting ecosystems including human beings [4], as stated by Singh and Walker [5] who suggested that these compounds possess high toxicity to mammals.

Extensive application of the insecticides led to soil contamination and therefore lowered soil fertility. The problem is accumulating due to deficiency of an applicable solution in the form of methods that are effective and ecologically friendly in the removal of toxic pollutants [6]. In general, physical and chemical clean-up technologies are relatively expensive, less sustainable and not appropriate. Boopathy [7] applied bioremediation on insecticides contaminated soil by using microorganisms, and the result was promising; but when bacteria was applied in degrading chlorpyrifos, its antibacterial metabolite

3,5,6-trichloro-2-pyridinol (TCP) prevented the proliferation of degrading bacteria [8], subsequently persisting in the environment.

The impediment of bioremediation on chlorpyrifos contaminated soil can be resolved by using fungi, as Chen et al. [9] demonstrated from the biodegradation of chlorpyrifos and TCP by fungus *Cladosporium cladosporioides* Hu-0. Abd El-Ghany and Masmali [10] showed a type of entomopathogenic fungus *Metarhizium anisopliae* (Met.) was able to biodegrade more than 90% of malathion. Ong et al [11] investigated the interaction between *M. anisopliae* and chlorpyrifos in the control of house fly, and found the residues of chlorpyrifos were significantly lower compared to the control in the potato-dextrose agar (PDA) culture media after 14 days. Therefore, this facilitated our interests on the possibility of *M. anisopliae* in biodegradation of chlorpyrifos in the common substrate-soil. In this study, we aimed to investigate the degradability of *M. anisopliae* on the chlorpyrifos and cypermethrin in soil by analysing the residues after certain incubation duration.

2. Materials and methods

2.1 Fungal Culture

Metarhizium anisopliae (Met.) Sorokin strain was isolated from the spores that form on the *Oryctes rhinoceros* (L.) beetle. The isolate was batch cultured on 10 Petri dishes containing potato-dextrose agar (PDA) at 27°C for 30 days. The conidial suspension prepared in the testing was harvested from young colonies from the surface of these colonies by using a sterilized L-rod and transferred aseptically to a tube containing a mixture of 0.1% (v/v) Tween 80 and autoclaved distilled water. The stock suspension was standardized at a concentration of 3.5 x 108 conidia ml-1 using a Neubauer haemocytometer.

2.2 Identification of fungus

Fungus conidia morphology was observed using compound microscope (Leica® DM500) with Köhler illumination. Samples were prepared as proposed by Humber [12], where the conidia were taken by an insect pin with assistance of a stereo microscope (Olympus SZ Stereo Microscope). The morphology was identified according to the Key to Fungal Entomopathogens [12], in which the *Metarhizium* was confirmed with its formation of short to long chains of conidia, with rounded to broadly conical apices, branched, densely intertwined conidiophores that forming a compact hymenium, and the conidia borne in parallel chains and green in mass (figure 1).

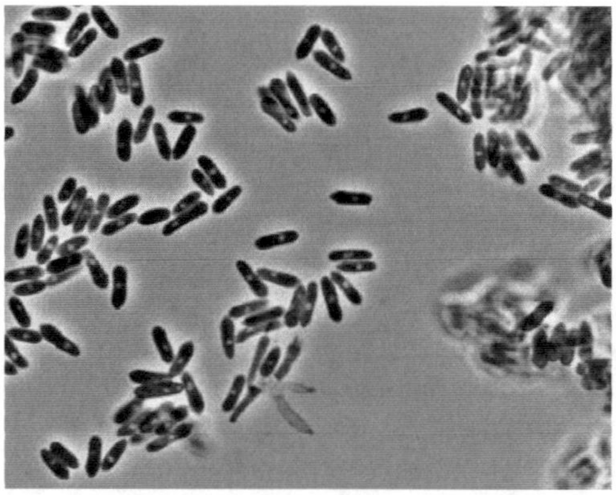

Figure 1. Conidia of *Metarhizium*.

2.3 *Degradability test on soil*

Soil samples were collected from the field with no history of insecticides application at Penang Bayan Lepas, Malaysia from a depth of 10-20 cm. Soil samples were sieved through a 90-mesh sieve to remove plant debris and stones. These soil samples were autoclaved and stored at 4°C. Ten grams of sterile soil were introduced in Petri discs, with the variants of three controls; 1) soil, 2) soil + insecticides, 3) soil + fungal inoculums, and the treatment; soil + insecticides + fungal inoculums in five replicates. Each of the insecticides was added at 500 ppm, and inoculations were done with two mL of fungal spore suspension (3.5 x 10^8 conidia ml^{-1}). The Petri discs were incubated at 30°C for 21 days then proceed for insecticidal residue analysis.

2.4 *Insecticidal residue detection*

A standard curve was initially generated to determine the particular retention time for each tested insecticide by preparing serial dilutions of five standards ranging in concentration from 50 to 500 ppm by weighing analytical grade chlorpyrifos (99.7%) and cypermethrin (94.3%) (Sigma-Aldrich, Malaysia) in a 25-ml volumetric flask. Acetonitrile (HPLC grade, Fisher Chemical) was used as the solvent. Modified solvent direct-immersion extraction- SDIE [13] was used to extract insecticide from the soil. Two grams of homogenized soil for treatments and controls were prior extracted using acetone, water, and an insecticide-specific solvent (chlorpyrifos-acetone and cypermethrin-hexane) in a 1:1:2 ratio with rotary shaking and immersed for 24 h. The upper solvent layer of the sample later was subjected to centrifugation using an Eppendorf Centrifuge 5427R (Eppendorf Asia Pacific Sdn. Bhd, Selangor, Malaysia) at 2,000 rpm for 5 min. The supernatant was cleaned-up using solid-phase extraction (SPE) C-18 (Supelclean ENVI-18 SPE wt. 500 mg, volume 6 ml) cartridge (EPA 1996). The filtrate was concentrated until dry at 55-65°C in a ventilated oven. The dried residues were reconstituted in 1 ml of acetonitrile in 2-ml amber glass vials for UPLC analysis.

The residues were analysed using the ACQUITY UPLC WATER system (Waters Analytical Instruments Sdn. Bhd., Petaling Jaya, Malaysia), consisting of a PU-1580 pump coupled to an HG-1580-31 mixer and a photodiode array (PDA) detector with programmable excitation and emission wavelengths. Separation was achieved using an ACQUITY UPLC BEH C18 Column (1.7 mm by 2.1mmby 100 mm). The PDA detector was set at the excitation wavelengths of 220 nm for chlorpyrifos and 225 nm for cypermethrin, with the initial mobile phase of 10:90 (v/v) for cypermethrin and methanol/acetonitrile at 70:30 (v/v) for chlorpyrifos. The quantitative measurement of the insecticide residue followed the CDFA standards [14].

3. Results and discussion

Extensive use of chlorpyrifos and cypermethrin may cause various environmental consequences due to the natural degradation, particularly soil fertility. We proposed the application of biodegradation on both chlorpyrifos and cypermethrin in agricultural soils that were also demonstrated by Abd El-Ghany and Masmali [10] that using *M. anisopliae* in reducing organophosphates in the soil. The principle to support the investigation of using fungi on chlorpyrifos and cypermethrin was due to the characteristics of the fungus that did not have sensitive targets for the insecticidal action mode of chlorpyrifos and cypermethrin [15].

The characteristic of fungus was identified as described in the section of 2.2 "Identification of fungus". The insecticides' residue for the controls (soil, soil + insecticides, soil + *M. anisopliae*) were less than 0.01 ppm, and the chlorpyrifos and cypermethrin's residue for the plates of soil+insecticides+*M. anisopliae* were significantly lower compared to the plate containing the soil with the two insecticides, respectively ($P < 0.05$, table 1). The result of this study was comparable to Ong et al. [11], in which the *M. anisopliae*+ChCy (a commercial insecticide that was having both chlorpyrifos and cypermethrin) showed significantly lower residues than the control. Our results agreed well with the study of Abd El-Ghany and Masmali [10] that showed *M. anisopliae* was able to biodegrade more than 90% of malathion, which had the same category and action mode as chlorpyrifos. Siewiera et al.

[16] used *Metarhizium robertsii* (genus *Metarhizium*) to enhance the tributyltin (TBT) degradation by estradiol (E2), in which the presence of *M. robertsii* significantly reduced the amount of tributyltin.

Table 1. Insecticide residues (ppm, means ± S.E) of the treatment soil after 21 days.

	Residue (ppm, means ± S.E)[#, *]	
	Chlorpyrifos	Cypermethrin
Soil	Nil	Nil
Soil + *M. anisopliae*	Nil	Nil
Soil + chlorpyrifos	262.6±7.60a	Nil
Soil + chlorpyrifos + *M. anisopliae*	19.39±0.10b	Nil
Soil + cypermethrin	Nil	194.4±4.30a
Soil + cypermethrin + *M. anisopliae*	Nil	19.68±0.36b

[#] Means ± S.E followed by same letter indicates not significantly different at P = 0 .05 within row, post hoc test, LSD

[*] Nil indicated that the residue is less than 0.01 ppm; S.E standard error

Chlorpyrifos is suggested as a high persistency pollutant in the environment due to its antibacterial metabolite 3,5,6-trichloro-2-pyridinol (TCP) [17]; however, Chen et al. [9] has shown the possibility of using the fungus *Cladosporium cladosporioides* Hu-01 to degrade chlorpyrifos and hydrolyzed its antibacterial metabolites 3,5,6-trichloro-2-pyridinol (TCP). Similarly, a study by Fang et al. [18] showed that application of the fungus *Verticillium sp.* from soil successfully degraded chlorpyrifos, Mukherjeea and Gopala [19] also demonstrated that two soil fungi *Aspergillus niger* and *Trichoderma viride* are able to degrade chlorpyrifos. The fungus may be an excellent alternative for bacteria in bioremediation of chlorpyrifos contaminated soil. Bioremediation on insecticides contaminated soil was achieved, as demonstrated by Chalamala et al. [20] who used *Aspergillus niger*, as an alternative to other conventional technique for the degradation of malathion contaminated residue soils in which *Aspergillus sp.* showed tolerance limit of 800 mg of malathion and degraded 300 mg within 24 h of incubation. The fungus may synthesize phosphotriesterases (PTE), the main class of enzymes in the hydrolysis of organophosphate insecticide such as chlorpyrifos. Various PTEs have been identified such as organophosphate hydrolase (OPH), methyl parathion hydrolase (MPH), organophosphorus acid anhydrolase (OPAA), diisopropylfluorophosphatase (DFP), and paraoxonase 1 (PON1), carboxylesterases from fungus [21]. In conclusion, future study could be broadened in the molecular extraction and catalytic enzymes from fungus on insecticides such as chlorpyrifos and cypermethrin, and also other organic insecticides.

4. References

[1] Díaz M F, Ramírez A., Poveda K. 2012. *Biol. Control* **60**, 182–191

[2] Casida J E. 2009. *Chem. Res. Toxicol.* **22**, 609–619.

[3] Racke K D, Coats J R. 1988. *J Agric Food Chem* **36**, 193-199.

[4] Liu Y H, Chung Y C, Xiong Y. 2001. *Environ Microbiol* **67**, 3746-3749.

[5] Singh B K, Walker A. 2006. *FEMS Microbiol Rev* **30**, 428-471.

[6] Schoefs O, Perrier M, Samson R. 2004. *Appli. Microbiol and Biotech.* **64**, 256-261

[7] Boopathy R. 2000. *Int Biodeterior Biodegradation* **46**, 29-36.

[8] Racke K D. 1993. *Rev. Environ. Contam. Toxicol.* 131-150.

[9] Chen S, Liu C, Peng, C., Liu, H., Hu, M. 2012. *PLoS ONE* 7: e47205.

[10] Abd El-Ghany T M, Masmali I A. 2016. *J Plant Pathol Microbiol* **7**, 349.

[11] Ong S Q, H Ahmad, A H Ab Majid and Z Jaal. 2017. *J Med Entomol*, **54**, 1626-32.

[12] Humber R A. 1997. Fungi: Identification. In Manual of Techniques in Insect Pathology (L.A. Lacey, ed.). *Academic Press: London*, 153-185.

[13] Ong S Q, A H Ab Majid and H Ahmad. 2016. *J Econ Entomol* **109**, 952-7

[14] CDFA (California Department of Food and Agriculture) 2011 Center for Analytical Chemistry Environmental Analysis. Determination of Bensulide and Imidacloprid in Surface Water EMON-SM-05-023

[15] (IRAC) Insecticide Resistance Action Committee 2015. Mode of Action Classification Brochure in Insecticide Resistance Management Fifth Edition. http://www.irac-online.org/

[16] Siewiera P, Bernat P, Rozalska S and Długonski J. 2015. *Int Biodeterior Biodegradation*, **104**, 258-263

[17] Watts M. 2012. Pesticide Action Network North America

[18] Fang H, Xiang Y Q, Hao Y J, Chu X Q, Pan X D. 2008. *Int Biodeterior Biodegradation*, **16**, 294-303

[19] Mukherjee I and Gopal M. 1996. *Toxicol Environ Chem* **57**, 145-151

[20] Chalamala R, Mitta M N, Muppala G P. 2012. *International Journal of Basic and Applied Chemical Sciences* **2**, 108-115.

[21] Bigley A N, Raushel F M. 2013. *Biochim Biophys Acta* **1834**, 443-453.

IOP Conf. Series: Earth and Environmental Science **380** (2019) 012015 doi:10.1088/1755-1315/380/1/012015

Study on Heath Forest Species *Melaleuca cajuputi* as Potential Natural Herbicides Agent to Suppress Weed Growth in Landscape Management

Rashidi Othman[1]*, Wan Masyitah Wan Daud[1], Razanah Ramya[1], Nursyafica Nadia Johari[1] and Zainul Mukrim Baharuddin[1]

[1]Department of Landscape Architecture, Kulliyyah of Architecture and Environmental Design (KAED), International Islamic University Malaysia, 53100 Kuala Lumpur

**Corresponding author: rashidi@iium.edu.my*

Abstract. Weeds are diverse in habitats throughout the world. In landscape management, weed is of major concern because they compete for growth factors with landscape plant of interest. Modern management relies on the importance of synthetic chemicals to control weeds as unwanted plants to reduce the weed management cost. The use of herbicides is not an appropriate tool for controlling development of some weeds but can create negative effects to the environment. Managing the natural and landscape resources are not only designing for good views and environment merely, but need to sustain all of living environment through consideration. Therefore, this study was aimed to explore *Melaleuca cajuputi*, heath forest species as potential natural herbicides agent in sustainable landscape management. Three assessments were conducted to determine their effectiveness towards suppressing weeds without degrading the environmental quality index. Assessment of allelochemical compounds in *M. cajuputi* demonstrated that these species have allelopathic potential as a natural herbicide.

1. Introduction

In 1967, the Weed Science Society of America defined a weed as a plant growing where it is not desired or known as an unwanted and undesirable plant that utilize on land biodiversity and water resource and may affect human welfare. Although they account for not more than 1% of the total plant species on earth, they cause great problems nevertheless to humankind by interfering in food production, economic stability, health, and welfare [1]. Weed plants are one of the main problems in global agriculture. They are a setback for the management and harvest of the given culture, by decreasing productivity which leads to an annual loss of approximately U$95 billion. In landscape management, weeds are of concern because of their competition with other plants for growth factors. The discovery of synthetic natural herbicides for weed control in the late 1940s made reduced tillage practices more feasible because tillage was not the only method of weed control [2]. In Malaysia, dominant weeds are broad-leaved weeds in rice field [3]. The annual expenditure is approximately about US$4.10 million on herbicides for rice alone which almost 7% of the total cost of herbicides in Malaysia. Weeds constantly compete with crop plants to cause a considerable loss in their productivity, and have been documented as serious plant pests since ancient times [4]. The application of allelochemicals as natural herbicides for weeds management tool has been reported [5], including their importance and interaction with the environment.

2. Materials and methods

2.1. Study site

The site selection is located within Jambu Bongkok Forest Reserve (370 ha) at Rantau Abang, Terengganu (4°55'N, 103°21'E) with altitude varying between 3-12 meters. Rantau Abang is a small village located 80 km from South of Kuala Terengganu. This site was chosen because it is one of the areas that is surrounded by the heath forest ecosystem and still have the traces of the forest. The heath forest ecosystem in this area is located near to the village and road systems. Even though there are wide area covered with heath forest, most of the ecosystem is already affected by human disturbance [6]. The sandy soil of the heath forest is often lacking in nutrients including nitrogen which impedes plant growth in these forests [7]. Over time, the sandy soil is covered by a thick layer of organic matter, allowing vegetation to develop. Heath forest is extremely fragile, once the organic layer is removed, no vegetation can regenerate because only sand remains. This heath forest was mainly dominated by *Melaleuca cajuputi* species. The study area is divided into three part of the forest: matured trees area, semi-matured trees and seedling areas as shown in figure 1.

i. Matured trees with trunk diameter between 80 to 130 cm

ii. Semi-matured trees trunk with diameter between 25 to 50 cm

iii. Young seedling with trunk diameter between 2 to 10 cm

Figure 1. Category of heath forest species based on trunk diameter at Rantau Abang, Terengganu.

2.2. Plant maceration extraction

Sequential alkaline extraction method was used to extract free and bound phenolic compounds from sodium hydroxide (NaOH) for 12h in the oven at 60°C. The alkaline extract was treated with hydrochloric acid (HCl) to reach pH 2.0, centrifuged and the supernatant extract was collected then re-extracted with ethyl acetate [8]. The final concentration was resuspended with methanol for further analysis by HPLC.

2.3. Determination of Total Phenolic Compound (TPC)

TPC determination was conducted by using the Folin-Ciocalteau assay as reported by [9] with slight modification. Quantification was performed with hydrolysed samples. Results were expressed as gallic acid equivalence (GAE) per gram dry weight sample using TECAN microplate reader.

2.4. HPLC analysis of Phenolic Acids Content

HPLC analyses were proceed with Agilent 1200 series (Agilent Technologies, Palo Alto, CA, USA) equipped with the binary pump, an autosampler and a diode array detector (DAD). The HPLC column was a reverse-phase Zorbax SB-C18 (Eclipse 100 × 2.1 mm, 1.8 μm). The temperature of the column was set at 25°C. For the analysis, two mobile phases were used consist of 1% formic acid in water/ acetonitrile 90:10 v/v (phase A) and acetonitrile (phase B). the solvent gradient used developed as follows: 0-40% solvent B (0-20 min), 40-60% solvent B (20-25 min); 60-100% (25-25.1 min), 100% solvent B (25.1-35 min) and 100-0% solvent B (35-35.1 min) at a flow rate of 0.4 ml/min and with detection at 280 nm throughout the gradient [10]. The phenolic acid contents were identified through their retention times and their UV spectra as compared to standards.

2.5. Determination of allelochemical effect through In Vitro model system

Seeds of the *Chloris barbata* (grass) were collected from a paddy field in Kepala Batas, Pulau Pinang. The tissue culture processed was modified based on [11]. The concentration of the water extractions was 100g/ L prepared for the experiment [12] from low to high concentration (0.5, 2.5 and 5.0 g/L). The experiment was observed during germination of weed species with three different stage of growth; Pre-emergent (Seed), Emergent (Seedling), Post-emergent (Individual plants). All stages of weed will be assessed with three different concentrations (high, medium and low concentration) at for four different periods of time (1st week – 4th week).

2.6. Statistical analysis

Data were expressed as the mean ± standard deviation of triplicates solvents extraction for total phenolic compound (TPC) and phenolic acids content. One-way analysis of variance (ANOVA) with Tukey's test was conducted using XLSTAT-Pro (2014) statistical software (Addinsoft, Paris, France).

3. Result

3.1. Analysis of phenolic acid content and composition in Melaleuca cajuputi

Three different stages of *M. cajuputi* vegetative samples were analysed for total and individual phenolic content to investigate the stability of their phenolic acid composition profile. Analysis of variance on the data exhibited highly significant differences between the three stages and the individual phenolic acid content (figure 2). This further reinforced different stages of *M. cajuputi* tree can have a marked influence on the accumulation of allelochemicals. The changes in phenolic acid composition are complex because it changes with different period of tree age as indicated by the interaction components. The total phenolic content for young, semi-matured and matured *M. cajuputi* tree ranged from 892.97 to 1376.28 µg/g DW. The semi-matured tree (1376.28 µg/g DW) was detected to have the highest amount of total phenolic content while the matured tree (892.87 µg/g DW) was the lowest.

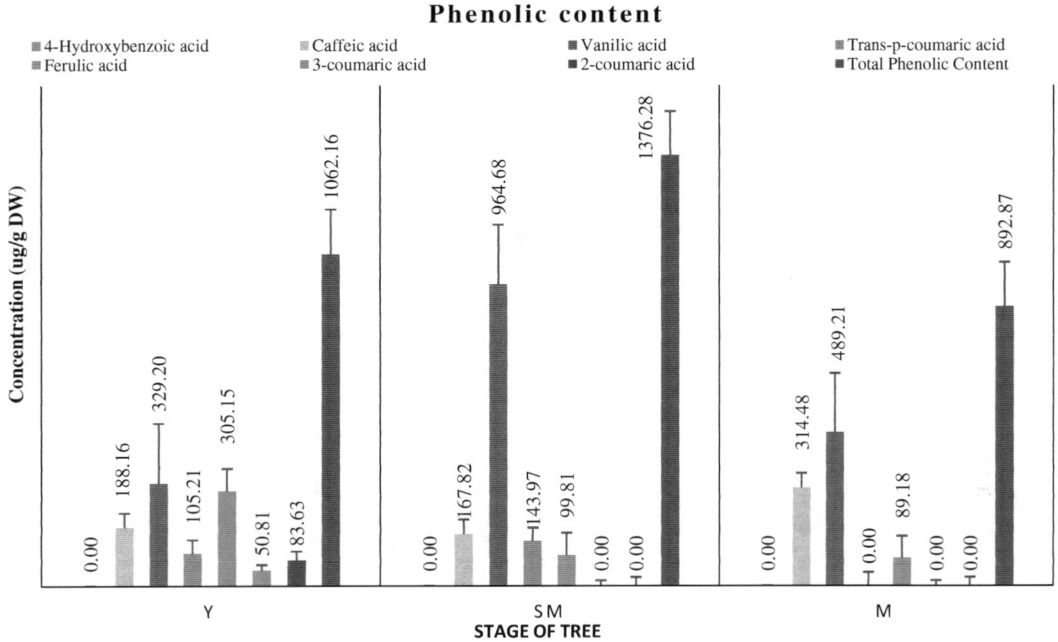

Figure 2: Comparison of phenolic acid content and composition of *M. cajuputi* between young, semi-matured and matured tree.

Comparison of phenolic acid profiles of the three different period of age showed variations in the profiles of individual phenolic acid compounds (figure 2). For example, young stage of tree with trunk diameter between 2 to 10 cm accumulated 6 major phenolic acid compounds mostly caffeic acid, vanillic acid, trans-p-coumaric acid, ferulic acid, 3-coumaric acid and 2-coumaric acid, whereas in semi-matured tree with trunk diameter between 25 to 50 cm accumulated only 4 types of phenolic acid compounds which are caffeic acid, vanillic acid, trans-p-coumaric acid and ferulic acid. For matured tree with trunk diameter between 80 to 130 cm, only three major phenolic acids were detected which are caffeic acid, vanillic acid and ferulic acid.

3.2. Analysis of allelopathic effect towards weed germination

All three stages of weed were assessed with three different concentrations (high, medium and low concentration) at four different periods of time (1st week – 4th week) (table 1).

Table 1. Effect of allelochemical extracts of *M. cajuputi* on the *C. barbata* germination and growth at different period of time at low, medium and high concentration.

Extract Concentration (g)	Shoot length (cm)	Shoot number	Radicle length (cm)	Plant height (cm)	Status
			WEEK 1		
Low (0.5)	0.4	55	0.2	1.2	Rapid growth
Medium (2.5)	0.4	5	0.1	1.1	Slow growth
High (5.0)	0.4	5	0.1	1.0	All radicle stunted
			WEEK 2		
Low (0.5)	0.5	110	0.3	1.4	Rapid growth
Medium (2.5)	0.4	5	0.1	1.1	2 radicle stunted
High (5.0)	0.3	5	0.1	1.0	All radicle stunted
			WEEK 3		
Low (0.5)	0.6	157	0.4	1.6	Rapid growth.
Medium (2.5)	0.4	5	0.1	1.1	3 radicle stunted
High (5.0)	0.3	5	0.1	1.0	All radicle stunted
			WEEK 4		
Low (0.5)	0.7	230	0.6	1.9	Rapid growth.
Medium (2.5)	0.4	5	0.1	1.1	All radicle stunted
High (5.0)	0.3	5	0.1	1.0	All radicle stunted

According to the analysis of allelopathic effect towards weed germination (growth rates of *C. barbata*) in this vitro model system, the seeds started germinating at day 2. Table 1 showed the effect of the allelochemical extracts from *M. cajuputi* on pre-emergent (seed), emergent (seedling), and post-emergent (individual plants) of *C. barbata* from week 1 until week 4. The result demonstrated that *C. barbata* seeds development showed a gradual inhibitory effect in germination rate through shoot number, radicle length, shoot length and plant height in all concentrations. The inhibition of seedling development increased from week 1 until week 2 as compared to low concentration of 0.5g/L. Concentration at 0.5 g/L was observed to have stimulatory effect on shoot length and number as well as radicle length and plant height. The number of seedlings of *C. barbata* showed a reduction of growth from 2.5 g/L until 5.0 g/L concentrations as compared to 0.5 g/L. In general, *M. cajuputi* allelochemical extract at 2.5 and 5.0 g/L showed constituents of allelopathic activity which influenced the inhibitory effect on pre-emergent, emergent and post-emergent of *C. barbata.*

4. Discussion

Both species studied at the cultivar levels was reported to change in different environments. [13, 14]. The bioavailability of phenolic acid compounds is a complex issue and depends on many factors such as location, times of year, cultivar and their relationship of all factors [15]. In this study, looking at the influences of different age of tree and total phenolic acid content interactions, variations in both the total phenolic content and the individual phenolic acid compounds exhibit strong relationships between *M. cajuputi* stage or period of maturity. Therefore, major factors influencing allelochemicals production are location, age and their interactions (figure 2). Genotype x environment interactions on biochemical composition has been previously reported for phenolics accumulation in potatoes. [16] reported that potatoes cultivated on loam soils in warm dry regions with low altitudes over a three-year period contained a lower amount of total phenolics than those cultivated in cooler and more humid regions on sandy loam soil. In a similar study they found that organically grown potatoes contained higher levels of phenolics than did the same varieties grown in a conventional manner. It is hypothesized that the chemically untreated plants defend themselves against unfavourable extrinsic factors with higher levels of polyphenols. Furthermore, it is known that plants that are exposed to abiotic and biotic stresses increase their production of phenolics as a defence mechanism [16, 17].

5. Conclusion

This research will establish a new body of knowledge on the application of allelopathic species of *Melaleuca cajuputi* as potential natural herbicides in landscape towards sustainable landscape management. Creation of bio herbicides based on allelochemicals generates the opportunity to exploit natural compounds in plant protection and shows the possibility to cope with evolved weed resistance to herbicides and may be based on the structure of compounds occurring in nature, which could be used without any risks as selective and eco-friendly herbicides. Eco-friendly trends in weed management drives scientists to reach for innovative sources and tools. A natural control strategy may be warranted because of concerns about environmental effects of chemical herbicides or mechanical control, lack of efficacy of other methods, or simple economics. Therefore, a natural control in weeds management today introduces a new finding towards bio herbicide products which are safer to be used as well as by using reactions between plants.

6. References

[1] Qasem J R and Foy C L 2001 *J. Crop. Prod.* **4** 43.
[2] Voltarelli V M, Pedro J, Ribeiro N and Lima M I S 2012 *Acta. Bot. Bras.* **26** 779.
[3] Karim S M R, Azmi M and Ismail B S 2004 *Weed. Biol. Manag.* **4** 177.
[4] Jabran K, Mahajan G, Sardana V and Chauhan B S 2015 *Crop. Prot.* **72** 57.
[5] Nornasuha Y and Ismail B S 2017 *Malays. Appl. Biol.* **46** 1.
[6] Syurhani A W, Hakeem K R, Faridah-Hanum I, Alias M S and Ozturk M 2014 *Emir. J. Food Agric.* **26** 1114.
[7] Rashidi O and Ruzaimi M R 2013 *Hutan heath: Khazanah landskap dan warisan ekologi Negara*. (Kuala Lumpur: Yamani Angle Sdn. Bhd.)
[8] Aarabi A, Mizani M, Honarvar M, Faghihian H and Gerami A 2016 *J. Food. Meas. Charact.* **10** 42.
[9] Baba S A and Malik S A 2015 *J. Taibah. Univ. Sci.* **9** 449.
[10] Zhao H, Sun J, Fan M, Fan L, Zhou L, Li Z and Guo D 2008 *J. Chromatogr.* A **1190** 157.
[11] Othman R 2009 *Biochemistry and genetics of carotenoid composition in potato tubers* (Ph.D. thesis. Lincoln University).
[12] Ismail B S, Syamimi H, Wan Juliana W A and Nornasuha Y 2016 *Sains. Malay.* **45** 517.
[13] Dencic S, Kastori R, Kobiljski B and Duggan B 2000 *Euphytica.* **113** 43.
[14] Banziger M and Cooper M 2001 *Euphytica.* **122** 503.
[15] Fraser P D and Bramley P M 2004 *Prog. Lipid Res.* **43** 228.
[16] Hamouz K, Lachman J, Vokal B and Pivec V 1999 *Rost. Vyroba.* **45** 293.
[17] Lewis C E, Walker J R L, Lancaster J E and Conner A J 1998 *Aust. J. Plant. Physiol.* **25** 915.

Acknowledgements
The research was supported by the Ministry of Higher Education Malaysia (MOHE) and International Islamic University Malaysia (IIUM) under research grant PRIGS18-001-0001.

IOP Conf. Series: Earth and Environmental Science **380** (2019) 012016 doi:10.1088/1755-1315/380/1/012016

Characterisation of allelochemical compounds signature in two mangrove forest species of *Rhizophora apiculata* and *Acrostichum aureum* and potential in suppressing weed growth

Rashidi Othman[1], Razanah Ramya[1], Norazian Mohd Hassan[2], Wan Masyitah Wan Daud[1] and Nursyafica Nadia Johari[1]

[1] International Institute for Halal Research and Training (INHART), Herbarium Unit, Department of Landscape Architecture, Kulliyyah of Architecture and Environmental Design (KAED), International Islamic University Malaysia, 53100 Kuala Lumpur
[2] Kulliyyah of Pharmacy, International Islamic University Malaysia, 25710 Kuantan, Pahang

**Corresponding author: rashidi@iium.edu.my*

Abstract. *Rhizophora apiculata* and *Acrostichum aureum* are a common mangrove species in Malaysia. Allelochemical interaction of the mangrove species was speculated to play an important role in dominance in the harsh mangrove environment. This study assessed the quantitative and qualitative determination for total phenolic content and individual phenolic compounds as allelochemical content for *R. apiculata* and *A. aureum* leaves as a potential allelopathic substance. The second objective is to investigate the allelopathic effect towards weed germination and growth through *in vitro* model system. Two types of maceration extraction comprising water extraction and sequential alkaline extract were used for allelochemical screening analysis. Both extractions were separated by ethyl acetate. *Acrostichum aureum* had the highest total phenolic content (1129.52 μg GAE/g DW) in comparison with *R. apiculata*. Meanwhile, the quantitative analysis by HPLC, revealed that different extraction detected different individual phenolic acid, Caffeic acid detected in both mangrove species in sequential alkaline extraction by *A. aureum*. The allelopathic effect of *Chloris barbata* for *A. aureum* extract indicated that during pre-emergent development, the suppression of weed growth was higher compared to post-emergent. Therefore, *A. aureum* may work as an allelochemical producer and can contribute to the establishment of pure colonies of *A. aureum* in the mangrove ecosystem.

1. Introduction

In the present day, a principal way of controlling weeds relies on herbicides [1] with the cost of using these products totalling in excess of 4.6 billion. About 3 million tonnes of herbicides are used to control weeds per year [2]. Many reports established on herbicides as alternative weed management but continuing use of herbicides in heavy doses of chemicals creates environmental pollution and increases the number of herbicide resistant weeds [2-5]. Approximately 99% of herbicides are released into the air, water and soil and only 1% reaches the targeted weed [6]. Such changes become crucial to preserve natural resources and product quality. Since triazine herbicides and photosynthesis inhibitors widely implemented with inappropriate application of herbicides, it could contribute the accumulation of active compound in the soil and weed species [7].

Based on the current statistic of natural product-based practices, almost 70% are registered active pesticide ingredients have their origins in natural products research however only 7% of biochemical approved by the USEPA are bioherbicides [8]. Due to that, exploiting the allelopathy of plant-plant interaction as an alternative to control weed evokes a new prospect [9-10]. In the last decades, there has been an increasing approach known as allelopathic or weed suppressive as a green alternative that produce secondary metabolite termed allelochemical [11-15].

The term allelopathy was introduced by Austrian plant physiologist, Hans Molisch in 1937 who stated that allelopathy means biochemical interaction among higher plants, and between higher plants and microorganism, involving stimulatory and inhibitory actions [16-17]. In addition, by using the term allelopathy the plants that contribute allelochemicals are known as the 'donor' plant and the influenced plant recognized as the 'target' plant [18]. Allelopathy can be used to combat the challenges of environmental pollution and herbicide resistance development, and significantly for ecological, sustainable, and integrated weed management systems [19]. Thus, allelopathy has great potential in the production of phytotoxins to suppress another plant through natural interaction, through changes in the physiological process of plant growth. Allelochemicals were detected through species distribution, production of toxin, phytotoxicity effect and physical factors [20]. A natural component such as glucosinolate is considered the precursors of the allelochemical processing phytotoxicity as an effective way to suppress weeds growth [21]. In allelopathy context, phenolic compounds are described as simple aromatic phenols, hydroxyl and substituted benzoic acids and aldehydes, hydroxy and substituted cinnamic acid, coumarin, tannin and flavonoids [22]. Phenolic acid is a subclass of the larger phenolic category that occurs in food plants (fruit, vegetable, and grains) as ester or glycosides conjugated with other natural compounds such as flavonoids and alcohols and contain a phenolic ring and at least one organic carboxylic acid function [23]. Phenolic compounds are one major class of identified allelochemicals present in many submerged aquatic plants [24].

Weeds are often the biggest threat to organic crop production due to their dynamic community [25]. Weed is a serious pest of rice and causes annual worldwide rice yield loss by weed is 15-21% in which 42 different weeds species were identified from different categories of annual, perennial, grassy weeds, sedges and broadleaved weeds [26]. This study focuses on grassy weeds *Chloris barbata* found in abundance at selected study sites. Mangrove forest is known as harsh environment with complex abiotic factors, and different species composition which result in mangrove habitat complexity. However, the role of phytotoxic allelochemicals in mangrove community is still unknown, with only limited availability of information on relationship between allelopathy and mangrove succession [27].

This study aimed to investigate two mangroves species namely *R. apiculata* and *A. aureum* as potential allelopathic extracts. Three objectives were developed to achieve the research aim which was to (i) assess total phenolic content in *R. apiculata* and *A. aureum* qualitatively and to (ii) assess individual phenolic compounds quantitatively as allelochemical content using maceration extraction methods (water extraction and alkaline sequential extraction), and to (iii) investigate the allelopathic effect of *R. apiculata* and *A. aureum* extracts towards weed germination and growth through *in vitro* model system on *Chloris barbata* pre and post emergent.

2. Materials and methods

2.1. Plant material and sample preparation

The selection of *R. apiculata* and *A. aureum* was done randomly. *R. apiculata* (bakau minyak) and *A. aureum* (piai raya) were collected at Bagan Lalang, Selangor (2°36'38.6"N 101°41'06.8"E) in the mangrove forest. About 3 kg of fresh *R. apiculata* and *A. aureum* leaves were collected randomly, and later freeze-dried for three days and pulverized to powder, then stored in a 4°C chiller at Herbarium Laboratory, IIUM.

2.2. Plant maceration extraction
Two types of extraction methods were applied namely water extraction and sequential alkaline extraction.

Water extraction. 10 g of powdered freeze-dried material was mixed with 100 ml of distilled water and agitated on incubator shaker for 30 minutes at room temperature. Then, the sample was incubated in the oven at 60°C for 30 min before allowed to stand overnight in darkness at room temperature. The following day the sample was filtered before the sample was re-extracted with ethyl acetate as detailed by [28].

Sequential alkaline extraction. Sequential alkaline extraction is a method used to extract free and bound phenolic compounds from sodium hydroxide (NaOH) for 12h in the oven at 60°C. The alkaline extract was treated with hydrochloric acid (HCl) to reach pH 2, centrifuged and the supernatant extract was collected then re-extracted with ethyl acetate [29]. The final concentration was resuspended with methanol for further analysis by HPLC.

2.3. Determination of Total Phenolic Compound (TPC)
TPC determination using the Folin-Ciocalteau assay as reported by [30] was used with slight modification. Quantification was performed with hydrolyzed samples. Results were expressed as gallic acid equivalence (GAE) per gram dry weight sample using TECAN microplate reader.

2.4. Determination of Phenolic Acids Content with High-Performance Liquid Chromatography (HPLC)
HPLC analyses were conducted with the Agilent 1200 series (Agilent Technologies, Palo Alto, CA, USA) equipped with the binary pump, an autosampler and a diode array detector (DAD). The HPLC column was a reverse-phase Zorbax SB-C_{18} (Eclipse 100 × 2.1 mm, 1.8 µm). The temperature of the column was set at 25°C. For the analysis, two mobile phases were used consist of 1% formic acid in water/ acetonitrile 90:10 v/v (phase A) and acetonitrile (phase B). the solvent gradient used were developed as follows: 0-40% solvent B (0-20 min), 40-60% solvent B (20-25 min); 60-100% (25-25.1 min), 100% solvent B (25.1-35 min) and 100-0% solvent B (35-35.1 min) at a flow rate of 0.4 ml/min and with detection at 280 nm throughout the gradient [31]. The phenolic acids content was identified through their retention times and their UV spectra as compared to standards.

2.5. Determination of allelochemical effect towards weed germination
Two allelopathic species selection namely; *R. apiculata* and *A. aureum* extraction were explained in section 2.2.1. Seeds of the *Chloris barbata* (grass) were collected from a paddy field in Kepala Batas, Penang. The tissue culture processed was modified based on [32]. The concentration of the water extractions was 1g/ L prepared for the experiment [33] from low to high concentration (10 g, 20 g, 30 g, 40 g and 50 g). The experiment was observed from week 1 until week 4 and average of weeds growth were measured during pre-emergent and post-emergent of *Chloris barbata.*

2.6. Statistical analysis
Data were expressed as the mean ± standard deviation of triplicates solvents extraction for total phenolic compound (TPC) and phenolic acids content. One-way analysis of variance (ANOVA) with Tukey's test was conducted using XLSTAT-Pro (2014) statistical software (Addinsoft, Paris, France).

3. Results

3.1. Analysis of total phenolic content in A. aureum and R. apiculata
These mangrove species showed significant differences in total phenolic content (TPC) ($p < 0.0001$; F:180.24; df: 10). *A. aureum* had the highest TPC between both mangrove species studied at 1129.52 (µg GAE/g DW) as presented in table 1.

Table 1. Total phenolic content (mg GAE/g DW) in two mangrove species.

Allelopathic species	Total phenolic content (µg GAE/g DW)
Acrostichum aureum	1129.52 ± 10.70
Rhizophora apiculata	977.90 ± 2.13

3.2. Analysis of phenolic acids content in A. aureum and R. apiculata

The phenolic acids analysis performed by HPLC system detected seven major phenolic acids: 4-hydroxybenzoic acid, caffeic acid, vanillic acid, trans-p-coumaric acid, ferulic acid, 3-coumaric acid and 2-coumaric acid. Analysis of variance established that both mangrove species exhibited highly significant differences in individual phenolic acids in water extraction ($p < 0.0001$). Both maceration extraction for phenolic acids content in two mangrove species as allelochemical is shown in figure 1. For water extraction, four individual phenolic acids were detected in *R. apiculata* (caffeic acid, vanillic acid, *trans-p*-coumaric acid and ferulic acid) whereas only three phenolic acids were detected in *A. aureum* (4-hydroxybenzoic acid, caffeic acid and 3-coumaric acid) (figure 1). The highest individual phenolic acid was ferulic acid at 1.56 µg/g DW in *R. apiculata*. Both mangrove species showed a different range of phenolic acid which was 0.02-1.56 µg/g DW in *R. apiculata* and 0.18-0.51 µg/g DW in *A. aureum*.

As for sequential alkaline extraction, the result showed that four individual phenolic acids were detected in *A. aureum* (4-hydroxybenzoic acid, vanillic acid, *trans-p*-coumaric acid and ferulic acid). However, only one phenolic acid was detected in *R. apiculata* (caffeic acid). The highest individual phenolic acid content was vanillic acid at 5.94 µg/g DW in *A. aureum.* In addition, the range of phenolic acid content for both mangrove species differed where 2.13-5.94 µg/g DW in *A. aureum* while 0.00-2.09 µg/g DW in *R. apiculata*.

Different extraction methods detected different individual phenolic acid. Sequential alkaline extraction produce the highest phenolic acids in both mangrove species compared to water extraction. *A. aureum* extract contained high amount of individual phenolic acid compared to *R. apiculata.*

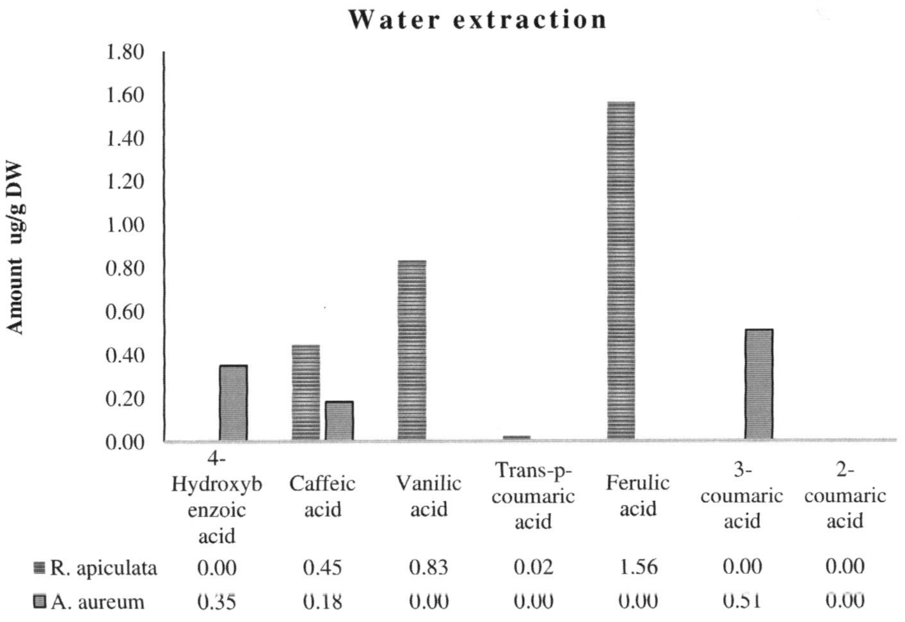

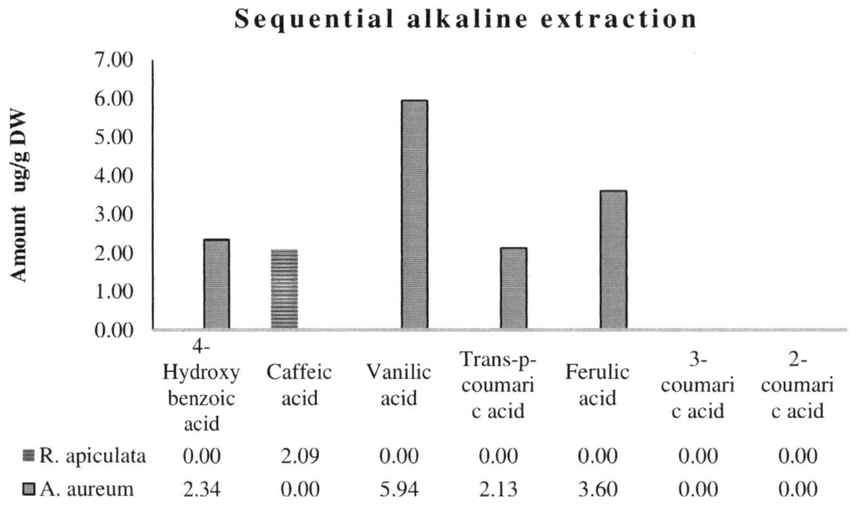

Figure 1. Comparison of individual phenolic acid content between *R. apiculata* and *A. aureum* using different method of extraction: water extraction (above) and sequential alkaline extraction (below).

3.3. Analysis of allelopathic effect towards weed germination

The purpose of this *in vitro* model system was to observe the propagation development of *C. barbata* seedling growth. Seeds of *C. barbata* were evaluated from day 0 until day 7 through *in vitro* model system cultured on MS basal medium as presented in table 2. From the observation, the *C. barbata* seeds began to germinate and grow in day 2 whereas shoot length, shoot number, radicle length, radicle number and plant height also increased. Table 3 showed the effect of the allelochemical extracts from *A. aureum* on pre-emergence of *C. barbata* seed after 7 days. *C. barbata* seeds development showed a gradual stimulatory effect in germination rate, radicle length, radicle number, shoot length and shoot number in all concentrations. The inhibition of seedling development decreased from 95% to 20% as compared to control except for 50 g/L. Table 4 presented that the effect of an allelochemical extract of the *A. aureum* showed progressive inhibition effects on seedling growth during post-emergence of *C. barbata* as the concentration increased. The number of seedlings of *C. barbata* showed a reduction in growth from 10 g/L until 50 g/L concentrations in all allelochemical extracts as compared to control. In general, *A. aureum* allelochemical extract on pre-emergence of *C. barbata* caused a reduction in shoot length, radicle length, plant height and a number of the seedling. To sum up, *A. aureum* showed constituents of allelopathic activity which influenced the inhibitory effect of *C. barbata* allelochemical extracts as well as concentrations.

Table 2. Seedling rate and growth index of *C. barbata*.

Day	Shoot height (mm)	No. of shoot	Length of radicle (mm)	No. of radicle	No. of seedling
Day 0	0.0±0.0	0.0±0.0	0.0±0.0	0.0±0.0	0
Day 1	0.0±0.0	0.0±0.0	0.0±0.0	0.0±0.0	0
Day 2	1.0±0.0	1.0±0.0	1.0±0.0	1.0±0.0	1
Day 3	3.8±0.2	1.0±0.0	7.7±0.5	0.9±0.3	26
Day 4	5.6±0.2	1.0±0.0	11.7±0.4	1.0±0.0	36
Day 5	7.0±0.2	1.0±0.0	12.6±0.5	1.0±0.2	37
Day 6	7.7±0.3	1.0±0.2	13.3±0.3	1.0±0.0	32
Day 7	8.7±0.3	1.3±0.5	13.4±0.4	1.0±0.0	31

Table 3. Effect of allelochemical extracts of *A. aureum* on the pre-emergent of *C. barbata* seeds after day 7.

Extract concentration (g/L)	Length of shoot (mm)	No. of shoot	Length of radicle (mm)	No. of radicle	No. of seedlings
0 (control)	9.9±0.4	1.4±0.5	10.9±0.4	1.0±0.0	25
10	9.5±0.4	1.0±0.0	8.5±0.5	1.0±0.0	2
20	14.5±0.1	1.0±0.0	8.5±0.8	1.0±0.0	2
30	5.0±0.0	1.0±0.0	5.0±0.6	1.0±0.0	3
40	9.0±0.4	1.0±0.0	8.5±0.9	1.0±0.0	2
50	10.0±0.0	1.0±0.0	7.0±0.0	1.0±0.0	1

Table 4. Effect of allelochemical extracts of *A. aureum* on the post-emergent of *C. barbata* seeds after day week 4

Extract concentrati on (g/L)	Length of shoot (mm)	No. of shoot	Length of radicle (mm)	No. of radicle	No. of seedlings
0 (control)	12.0±0.3	1.0±0.0	15.6±0.6	1.0±0.0	25
10	10.4±0.3	1.0±0.0	14.1±0.3	1.0±0.0	14
20	12.3±0.2	1.1±0.3	14.0±0.4	1.0±0.0	9
30	13.8±0.4	1.4±0.5	16.8±0.1	1.0±0.0	5
40	14.5±0.2	1.2±0.4	19.2±0.4	1.0±0.0	6
50	12.8±0.1	3.6±0.5	35.6±4.9	1.0±0.0	5

4. Discussion

Phenolic compounds is produced by the plant in response to environmental stress in the mangrove ecosystem. Fern species *A. aureum* is a mangrove species with tolerance to high salinity and environmental stress, hence high TPC was detected from this fern [34]. High salinity in the soil causes difficulty to absorb sufficient amounts of water, thus decreasing cell osmotic potential [35-36]. Furthermore, plants respond towards environmental condition by developing sophisticated mechanism such as cell regulation [37]. As stated by [38], allelopathic plants depend on resistance mechanism leading to detoxication: conjugation, sequestration and oxidation which may switch on partially to oxidative stress [34]. In this research sequential alkaline extract was found to be a potential resource of the bioactive allelochemical compound compared to water extraction. This finding was supported by [38] who mentioned that sequential alkaline extraction is a method used to extract free and bound phenolic compounds from plant materials. Phenolics in plant species largely appear to be linked with plant cell wall polysaccharides by both hydrophilic and hydrophobic bonds in which enzymes might disintegrate the phenolic-cell wall matrix bonds and enhance phenolic extraction as well as important for the stability of the phenolics in the extract [39].

The allelochemical extracts from *A. aureum* affected germination and growth differently. The result indicated that most of the target species demonstrated a gradual reduction of growth development at different inhibitions as concentrations increased. According to [40], plants have different abilities to tolerate allelochemicals to reduce the uptake of the allelochemicals also to detoxify the allelochemicals from the target site. According to [41], the exact mechanism involved on inhibitory effect was caused by allelochemicals interfering with the physiological and biochemical process in target species.

5. Conclusion

In conclusion, by considering the result from allelopathic treatment through the *in vitro* model system, the possible allelopathic mode of action and mechanism of the plant extracts that contained water-soluble (aqueous) phytotoxic constituents, as a potential natural herbicide was capable of interfering directly on germination and growth of other species especially weed. *A. aureum* and *R. apiculata* may produce phytotoxicity known as allelochemical to develop their community through regulatory, constitutive and inductive mechanism towards the environment as a physical defense system. In other words, it could be incorporated as a natural herbicide for weed management control as an organic tool in horticulture as well as landscape management.

6. References

[1] Hong N H, Xuan T D and Eiji T 2004 *Crop. Prot.* **3** 255.
[2] Bhadoria P B S 2011 *Am. J. Exp. Agric.* **1** 7.
[3] Duke S O and Cerdeira A L 2005 *Int. Centre for Genetic Engineering and Biotechnology (Trieste, Italy)* vol 2 pp 67–143.
[4] Setia N, Batish D R, Singh H P and Kohli R K 2007 *J. Environ. Biol.* **28** 63.
[5] Yap S Y and Ong H T 1990 *Environmentalist.* **10** 189.
[6] Botelho R G and Cury J P 2012 Herbicides and the Aquatic Environment Herbicides – Properties, Synthesis and Control of Weeds ed M N Hasaneen (Croatia: INTECH) chapter 9 pp 149–164.
[7] Soltys D, Krasuska U, Bogatek R and Gniazdowska A 2013 Allelochemicals as Bioherbicides — Present and Perspectives. Herbicides - Current Research and Case Studies in Use ed A Price (Croatia: INTECH) chapter 20 pp 517–542.
[8] Cantrell C L, Dayan F E and Duke S O 2012 *J. Nat. Prod.* **75** 1231.
[9] Kato-Noguchi H, Hamada N, Morita M and Suenaga K 2013 *J. Plant. Physiol.* **170** 1.
[10] Duke S O, William R D and Mankhart A H 1986 *Weed. Sci.* **35** 499.
[11] Dayan F, Romagni J, Tellez M, Romando A and Duke S 1999 *Pestic. Outlook.* **10** 185.
[12] Voltarelli V M, Pedro J, Ribeiro N, Inês M and Lima S 2012 *Acta. Bot. Bras.* **26** 779.
[13] Durán-Serantes B, González L and Reigosa M J 2002 *Acta. Physiol. Plant.* **24** 385.
[14] Bhowmik P C and Inderjit 2003 *Crop. Prot.* **22** 661.
[15] Bich T T N and Kato-Noguchi H 2012 *Aquat. Bot.* **103** 30.
[16] Inderjit and Duke S O 2003 *Planta.* **217** 529.
[17] Kohli R K, Batish D R, Singh H P and Dogra K S 2006 *Biol. Invasions.* **8** 1501.
[18] Rice E L 1984 *Allelopathy* (Florida: Academic Press)
[19] Jabran K, Mahajan G, Sardana V and Chauhan B S 2015 *Crop. Prot.* **72** 57.
[20] Dayan F E, Owens D K and Duke S O 2012 *Pest. Manag. Sci.* **68** 519.
[21] Putnam A R and Tang C S 1986 *The Science of Allelopathy* (New York: John Wiley and Sons)
[22] Petersen J, Belz R, Walker F and Hurle K 2001 *Agron. J.* **93** 37.
[23] Tsao R 2010 *Nutrients.* **2** 1231.
[24] Gao Y N, Liu B Y, Xu D, Zhou Q H, Hu C Y, Ge F J, Zhang L P and Wu Z B 2011 *Pol. J. Environ. Stud.* **20** 1153.
[25] Clarka M S, Ferrisb H, Klonskyc K, Laninid W T, Van Bruggene A H C F and Zalomf G 1998 *Agric. Ecosyst. Environ.* **68** 51.
[26] Hakim M A, Juraimi A S, Ismail M R, Hanafi M M and Selamat A 2010 *J. Agron.* **9** 9.
[27] Chen L, Peng S, Chen B, Li J and Pang J 2009 *Allelopathy. J.* **23** 469.
[28] Tsatsaroni E 1998 *Dyes. Pigments.* **37** 307.
[29] Aarabi A, Mizani M, Honarvar M, Faghihian H and Gerami A 2016 *J. Food. Meas. Charact.* **10** 42.
[30] Baba S A and Malik S A 2015 *J. Taibah. Univ. Sci.* **9** 449.
[31] Zhao H, Sun J, Fan M, Fan L, Zhou L, Li Z and Guo D 2008 *J. Chromatogr.* A **1190** 157.
[32] Othman R 2009 *Biochemistry and genetics of carotenoid composition in potato tubers* (Ph.D. thesis. Lincoln University).
[33] Ismail B S, Syamimi H, Wan Juliana W A and Nornasuha Y 2016 *Sains. Malay.* **45** 517.
[34] Lai H and Lim Y 2011 *3rd. Int. Conf. on Chemical, Biological and Environmental Engineering (Singapore)* vol 20 pp 8–12.
[35] Duque A S, de Almeida A M, da Silva A B, da Silva J M, Farinha A P, Santos D, Fevereiro P and de Sousa Araujo S 2013 Abiotic Stress Responses in Plants : Unraveling the Complexity of Genes and Networks to Survive Herbicides and the Aquatic Environment Abiotic Stress-

Plant responses and applications in agriculture ed K Vahdati and C Leslie (Croatia: INTECH) chapter 3 pp 49–101.
[36] Gniazdowska A and Bogatek R 2005 *Acta. Physiol. Plant.* **27** 395.
[37] Inglett G E, Rose D J, Chen D, Stevenson D G and Biswas A 2010 *Food. Chem.* **119** 1216.
[38] Khoddami A, Wilkes M A and Roberts T H 2013 *Molecules.* **18** 2328.
[39] Zahed N, Hosni K, Ben Brahim N, Kallel M and Sebei H 2010 *Acta. Physiol. Plant.* **32** 1221.
[40] Yang Q, Ye W, Liao F and Yin X 2005 *Chin. J. Plant Ecol.* **24** 459.
[41] Santos P C, Santos V H M, Mecina G F, Andrade A R, Fegueiredo P A, Moraes V M O, Silva L P and Silva R M G 2015 *S. Afr. J. Bot.* **100** 114.

Acknowledgements
The research was supported by the Ministry of Higher Education Malaysia (MOHE) and International Islamic University Malaysia (IIUM) under research grant PRIGS18-001-0001.

IOP Conf. Series: Earth and Environmental Science **380** (2019) 012017 doi:10.1088/1755-1315/380/1/012017

Soil Water Analysis Tools (SWAT) hydrology modelling as a basis for spatial planning: a case study in Cimandiri Watershed, West Java Province

I Ridwansyah[1*], M Yulianti[1] and H Wibowo[1]

[1]Research Center for Limnology, Indonesian Institute of Sciences (LIPI), Cibinong Science Center-Botanical Garden, Cibinong 16911, Indonesia

**Corresponding author: iwanr@limnologi.lipi.go.id*

Abstract. Appropriate spatial planning can be analysed by looking at its impact on water resources on a watershed scale through a hydrological modelling approach. The objective of this study was to examine the effect of land-use change and the spatial plan of the region in the Cimandiri watershed using Soil Water Analysis Tools (SWAT) hydrological modelling. The Cimandiri watershed (1,830 km^2) is situated in West Java Province, Indonesia. This watershed has a strategic position because of its proximity to the Capital City of Jakarta. This study highlights the importance of hydrological response considerations in spatial planning. From 1994 to 2016, there has been a reduction of forest land cover (157.8 ha), paddy fields (688.5 ha) and plantations (216.4 ha). This land-use change affects the hydrological conditions of the Cimandiri Watershed. The results indicated that there was an increase in the surface flow of 8.7 mm/year (186.6 to 195.3 mm/year), from 1994 to 2016 while baseflow decreased from 781.64 to 777.7 mm/year. The verification models were in satisfactory categories (R2 0.62 and NSE 0.51). Therefore, this indicated hydrological conditions in 1994 were relatively better than in 2016. The simulation result on the spatial plan shows the surface runoff, and the base rate was 284.9 mm/year and 684.7 mm/year, respectively. Interestingly, land use proportion based on urban spatial planning generated relatively higher surface runoff and sediment yield compared to existing land use. Hence there should be a review of the existing urban spatial planning of Sukabumi Regency that was resulting in the worst hydrological response based on SWAT model.

1. Introduction

Spatial planning is a method and approach used by a government in regulating the spatial distribution of human activities that aims to improve the welfare of society [1,2,3]. Spatial planning provides essential information for a policymaker to determine the proper strategies in the management of natural resources [4,5]. The latest regulation on spatial planning in Indonesia is set in the Spatial Planning Law No. 26/2007. This regulation is then followed by several rules at the lower level. The law stated that spatial use refers to the spatial function specified in the spatial plan implemented by developing land, water, and air, as well as other natural resources.

Spatial planning cannot be separated from hydrological response within watershed scale. The water system must be adapted to the planning system in the region. Continued pressure on watersheds due to economic growth and population should also be considered in the development and preparation of the spatial plan in the region. One of the main factors affecting the hydrological cycle in the watershed is the anthropogenic factor, such as land-use change and land management [6,7]. It is clearly stated in the law that spatial planning needs to consider sustainability of water resources. However, the creation process of the spatial plan often does not take into account land uses change issue and its relationship to hydrology response [4] that has the potential to cause an imbalance of water resources. This process has implications for the development of a complicated and complex water resource system. The influence of land-use change on water balance in the watershed scale is a priority in the hydrological study because such change influences land-cover type that alters surface runoff generation and then affects the catchment hydrological process [7]. Therefore, the decision planning purpose of the region requires the fulfilment of water resources services.

One of the watersheds that need attention is Cimandiri Watershed, West Java, Indonesia. Based on reports from the West Java natural resource management centre, Cimandiri Watershed is one of the watersheds experiencing enormous land degradation. Cimandiri watershed is characterised by high sedimentation, extreme river fluctuation and increased pollution of river water. This information can be proven from the results of simulated data around the Cimandiri estuary, which showed that degradation has occurred from 1972 – 2015 (figure 1). Unwise land conversion is the main issue in Cimandiri Watershed that has an impact on severe erosion and high discharge. Moreover, mining activities such as gold, iron, and quartz cause sedimentation, abrasion, and coastal damage that adversely affect watershed ecosystems. Cimandiri Watershed is located in the Sukabumi region, which has the potential to experience rapid regional development. The construction of toll roads access to Sukabumi will accelerate land-use changes, thus increased the pressure on the watershed. Therefore, appropriate spatial planning based on land-use changes approach is needed to minimize the impact of regional development on the watershed ecosystem.

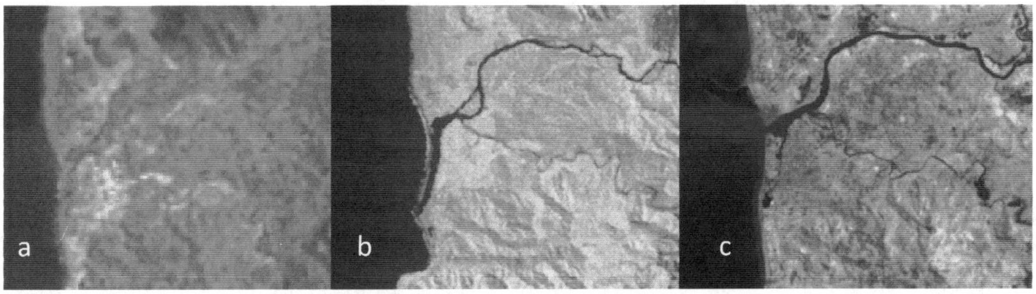

Figure 1. Change of Cimandiri Wateshed based on simulated data in year (a) 1972 (b) 1991, and (c) 2015.

Hydrological modelling is one of the tools that can be used to examine the suitability of spatial planning that has been applied. Among the most commonly used simulation models in spatial planning analysis is the Soil Water Analysis Tools (SWAT) model. SWAT is a GIS-based hydrologic simulation model developed by USDA Agricultural Research Service [8]. In recent years, the SWAT Hydrology Model has been widely used in watershed management as a hydrological model [9,10].

The objectives of this study was to: (1) analyse land-use changes in Cimandiri Watershed, (2) calculate the water balance of watershed under different land-use scenarios using the SWAT hydrology model. This information is significant to assist policymakers for spatial planning in Cimandiri Watershed as part of Sukabumi Regency.

IOP Conf. Series: Earth and Environmental Science **380** (2019) 012017 doi:10.1088/1755-1315/380/1/012017

2. Materials and methods

2.1. Study area

The study area was Cimandiri Watershed, situated in Sukabumi, which constituted a small part of Cianjur Regency, the southern part of West Java, Indonesia (figure 2). The catchment has an area of 4,128 km^2 and lies between latitude -6.701 to -7.175 and longitude 106.476 to 107.093. The average annual precipitation was 2,524 mm, and the average maximum and minimum temperature were 28°C and 20°C, respectively. The topography of the Sukabumi Regency, in general, includes a curved surface in the south and mountainous in the northern and central regions; and the height ranges from 0 - 2,960 m above sea level (asl).

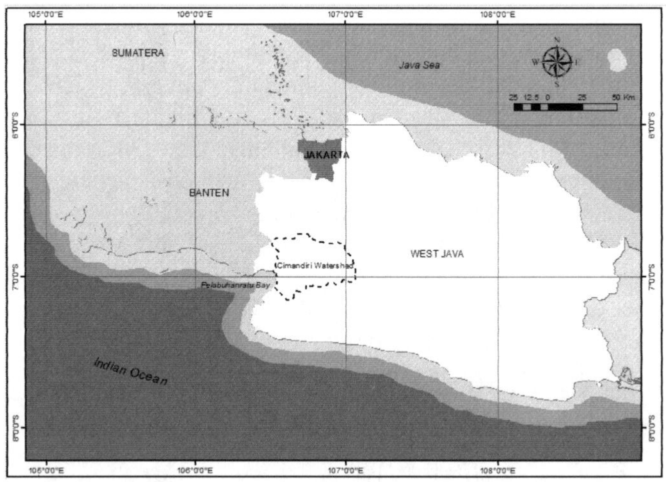

Figure 2. Change of Cimandiri Wateshed based on simulated data in year (a) 1972 (b) 1991, and (c) 2015.

2.2. Data collection

The data used in this study consisted of primary data collected from field sampling and secondary data in spatial and tabular forms derived from various sources (table 1). The collected data were used for two main stages of the study, which were land use analysis and hydrological modelling using SWAT.

Table 1. Summary of data requirements.

No	Margin	Data format	Source
1	Digital elevation model	Grid (cell size 30 x 30 m)	SRTM, USGS
2	Soil types scale 1:100.000	Vector (shapefile)	Land and agro-climate research centre, Ministry of agriculture
3	Meteorological data (daily) from two weather stations	Table (.dbf and text)	Meteorology and climatology agencies
4	Landsat image 1994		USGS
5	Landsat image 2016		USGS
6	Soil properties for SWAT database	Numeric	Field sampling and laboratory analysis
7	Daily discharge	Table (.dbf and text)	Ministry of public works

2.3. Land use classification
Land use classification in the Cimandiri Watershed was mapped by the use of Landsat imagery data from 1994 and 2016. Land use analysis was conducted with remote sensing methods and GIS. Land use analysis was initiated with the processing of satellite imagery as the primary data, followed by image processing that consists of two parts: pre-processing and image classification. Pre-processing of image includes a process of preparing image data for subsequent analysis which tries to correct or compensate for systematic errors, such as atmospheric, radiometric and spatial error, this process was for Landsat image 5 (the year 1994). The image classification method is used supervised classification, and the training sample area is derived from ground check data was collected by Camera GPS.

Accuracy of assessment is defined as a comparison of a map produced from remotely sensed data with ground check data. How closely the new map produced from the remotely sensed data matches to the source map is an important analysis [11]. Kappa coefficients were used to estimate the similarity between the two maps. Values > 80% are categorized as the same, 60% -80% of high similarity, almost the same 40% - 60% and < 40% do not have a common [12].

Detection of land-use changes in two periods was carried out using the cross-tabulation method. Comparisons will involve observing changes in each pixel [13,14, 15]. Detection produces a matrix of transformation. This matrix provides information on land-use changes in the region, spatially and quantitatively, including gains and losses between 1994 and 2016.

2.4. Model description
The SWAT model was utilised to examine the impact of land use on water balance components and sediment yield in Cimandiri Watershed. SWAT is a physically-based, distributed hydrological model regarded as a versatile model that is used to integrate environmental processes to support more effective watershed management [16]. A more detailed explanation of SWAT is available in several publications [17, 8, 18].

SWAT analysis was generated by ArcSWAT tools in ArcGIS 10.1. SWAT model requires a series of stages including watershed delineation, Hydrology Response Units (HRUs) definition, database input, calibration and validation. The first step was to delineate Cimandiri Watershed using Digital Elevation Model (DEM) into 57 sub-catchments. A further step was a division into the smallest spatial unit of the model, which is HRUs based on land uses, soil types and slope characteristics. After input of meteorological data, SWAT model was executed and continued for calibration and validation stages. Daily discharge data from several gauging stations in the study area was used in the calibration and validation process. The calibration process of the SWAT model used a manual method where the parameters that determine the discharge output were changed and adjusted to the natural conditions of the Cimandiri Watershed. Some parameters that are sensitive to discharge variation were also changed by trial and error. Model calibration was performed using data from 2009, while validation was used in 2010.

The performance of SWAT model in simulating streamflow was evaluated using Nash-Sutcliffe efficiency [19] and the coefficient of determination [20]. The NSE statistical model is widely used to show the performance of a model because it can provide more accurate information about the value given. Moriasi et al. [21] divided the statistical value criteria for NSE as: $0.75 < NSE < 1.00$ (very good), $0.65 < NSE < 0.75$ (good), $0.5 < NSE < 0.65$ (satisfying) and $NSE \leq 0.50$ (unsatisfactory).

After calibration and validation, SWAT was applied to predict the water balance, and sediment yield for five land uses scenarios (table 2). These scenarios changed the HRU distribution in the model; however, delineation of sub-catchments remained the same. Therefore, the results can reflect the impact of land use on the hydrological response as a basis for spatial planning.

Table 2. Five land use scenarios for SWAT model simulation.

No	Land use	Explanation
1	Land uses 1994 (LU 1994)	Land use resulting from image data processing
2	Land uses 2016 (LU 2016)	Land use resulting from image data processing
3	Urban spatial planning (RTRW) of Sukabumi Regency (USP)	Spatial planning of Sukabumi Regency (2012-2032)
4	USP and Best Management Practices (USP + BMPs)	Combination of spatial planning and Best Management Practices
5	Optimum land use (OLU)	Land use based on Law No. 41, 1999 and Law No. 37, 2014

3. Results and discussion

3.1. Land use changes in Cimandiri Watershed

The nine categories of land use composition produced from Landsat images for 1994 and 2016 are displayed in table 3. The results of image interpretation and classification show that there has been a considerable change during the twelve years, even though the major land use did not change significantly. The major land-use types in Cimandiri Watershed were plantation and agriculture, accounting for 26 % and 24 % respectively. Nonetheless, these land uses were decreased. The decline also occurred in almost all natural land uses such as mixed forest, paddy fields and bushes. On the other hand, there is an increase in the built-up area, such as commercial area and settlement. Settlement increased by 1,044 ha (0.6 %) while commercial area improved by 137.4 ha (0.2 %). This result indicates that population growth and development program had been triggered by the expansion of settlement and commercial area. Based on data from West Java in Figure (Statistics Indonesia), the population in Sukabumi Regency rose from 1,832,966 to 2,434,221, or 32.80% between 1994 and 2016.

Table 3. Land use proportion of Cimandiri Watershed in 1994 and 2016.

No	Land uses class	Land uses 1994		Land uses 2016		Changes (ha)
		Area (ha)	Percentage (%)	Area (ha)	Percentage (%)	
1	Water bodies	980.5	0.5	999.7	0.5	19.2
2	Bushes	28,031.2	15.4	27,535.6	15.1	-495.6
3	Commercial area	280.9	0.2	418.3	0.2	137.4
4	Mixed forest	26,705.3	14.6	26,552.6	14.5	-152.7
5	Plantation	48,269.9	26.4	48,053.5	26.3	-216.4
6	Settlement	17,209.8	9.4	18,254.2	10.0	1,044.4
7	Open area	1,150.9	0.6	1,257.4	0.7	55.4
8	Paddy field	14,685.4	8.0	14,151.9	7.8	-533.5
9	Dry agriculture	45,251.5	24.78	45,342.2	24.8	-90.7
	Total	182,565.4	100.0	182,565.4	100.0	

3.2. Calibration and validation model

Figure 3 shows a comparison of daily hydrograph between observed and simulated streamflow in a one-year period. There was some period that simulated streamflow (blue line) that was outperformed compared to observed data (black dashed line). Simulated data was underestimated at the end of January and the beginning of August, and overestimated in May. However, SWAT model produced a relatively similar pattern of peak and low flows from observed daily hydrograph. The average streamflow simulated by SWAT (22.0) was close to the average of the observations (22.3).

Simulated and observed daily discharge matched well in calibration process with ENS = 0.62 and R2 = 0.786 and for validation ENS = 0.58 and R2 = 0.618 (figure 3). Even though the values were relatively low, they were still acceptable for the validation process [21]. The results of this calibration were almost similar to other work conducted using SWAT in the Cisadane watershed [22]. It is indicated that the SWAT model can be used in Cimandiri Watershed.

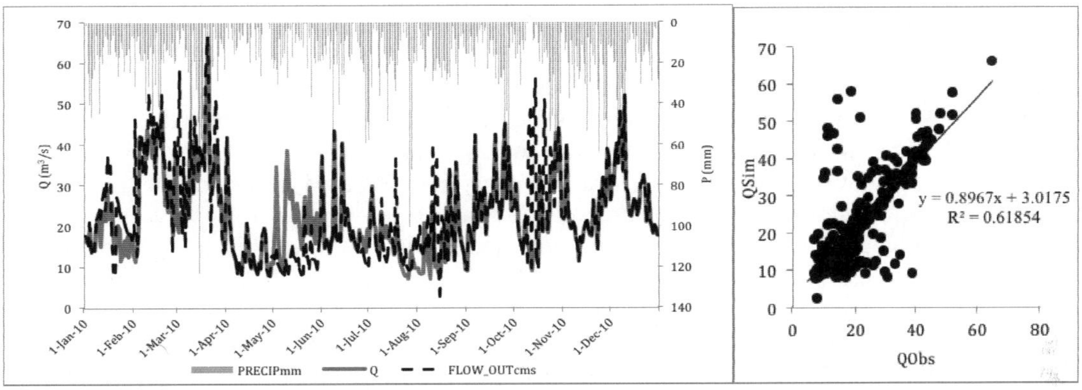

Figure 3. Comparison of observation and simulation streamflow.

3.3. Water balance hydrology model

Table 4 provides the respective results of simulation and comparative impact from different land-use scenarios in Cimandiri Watershed. The simulation results of SWAT model displayed a notable effect of varying land-use types of land use on hydrology response (table 4). It was found that the most significant consequence of land use was on the amount of surface runoff. The simulated of the annual hydrological response indicates that compared to land use in 1994, the annual surface runoff is 9 mm higher in 2016, increases of 4.82 %. The effects of land-use change was also directly linked to changes in other hydrological components such as interflow, baseflow and sediment load from the watershed. The impact of this change resulted from a decline in forest cover and a growth of the built-up area similar to results from previous studies in several parts over the world such as [23, 24, 25, 26, and 27].

Table 4. Comparison of model results for five scenarios.

Parameter	Land uses 1994	Land uses 2016	Urban Spatial Planning (USP)	USP + BMPs	Optimum land uses (OLU)
Surface runoff (mm/year)	186.58	195.27	284.9	220.2	170.4
Interflow (mm/year)	830.1	821.69	800.66	838.2	803.5
Baseflow (mm/year)	777.72	781.64	684.7	708	759.9
Sediment yield ($x10^3$ Ton/year)	1,227.7	1,233.6	2,505.7	2,119.6	1,061.7

Surprisingly, the simulation results indicate significant increase in annual surface runoff and sediment yield under land-use scenario using urban spatial planning of Sukabumi Regency for 2012 – 2032 (USP). Compared with the baseline land use (1994), there was 52.70% change for surface runoff and 104.10% for sediment yield. By adding conservation approach through Best Management Practices of agricultural land on USP (USP+BMPs), surface runoff and sediment yield were decreased, even though the values remain higher relative to other scenarios. The result of this simulation confirms previous studies that best management practices using terracing, contouring and ponds successfully reduce surface flow and sediment yield [28, 29, and 30]. However, BMPs application in the existing urban regional planning did not provide significant improvement of hydrological conditions proportionate to current land use. The best hydrological results were performed by simulation using the optimum land use (OLU). The OLU was based on government regulations which determine the proportion of protected forest areas in upstream and irrigated land in the region.

These findings further support the idea of taking into considerations of water resource management [31] and land-use change [32] in spatial planning. [33] argued that spatial planning based on land use approach is mitigation to prevent future risks on water balance caused by land-use change. Therefore, policymakers need to ensure that the impact of proposed land-use changes on water balance is considered during an early stage of spatial plan [34]. Furthermore, the planner should be aware that any possible adverse effects on the ecosystems will impact social and economic welfares, as well as other ecosystem services [2]. Ideally, spatial planning should be used to overcome problems regarding water availability through the utilization of soil and water conservation technology. Hence, the government, especially at the local government level as a spatial planner, must be able to balance the framework of land use planning with integrated water resources management.

The use of hydrological model potentially become one of the useful tools in spatial planning in a region. Integrating the effects of water resources into decision planning involves a robust model [35] to provide a better prediction of land-use impacts on a hydrologic response [36]. This study proved the potential of SWAT as one of the widely used hydrology models to support decision making in spatial planning. A physically-based SWAT model can be used to analyze the impact of land changes on hydrological responses in a watershed region. Furthermore, this model can assess the regional spatial plan, whether it has a good effect on sustainable water resources management. The findings of this study revealed the potential use of a hydrological model for a review of the regional spatial plan so that it can be implemented per the mandate of the law.

4. Conclusion

This study has shown that SWAT model is capable of simulating daily flow in Cimandiri Watershed within the range acceptable accuracy of model performance based on NSE value. It was also revealed that SWAT is an appropriate tool to be used as a basis for spatial planning. The findings of this study highlight the importance of taking into account the hydrological response to spatial planning. An implication of this is the possibility to re-examine the existing urban spatial planning of Sukabumi Regency that is resulting in the worst hydrological response compared to other scenarios.

5. References

[1] Leppert G, Hohfeld L, Lech M and Wencker T 2018 Impact, diffusion and scaling-up of a comprehensive land use planning approach in the Philippines: from development cooperation to national policies Bonn:Deutsches Evaluierungsinstitut der Entwicklungszusammenarbeit (DEval)

[2] Granit J, Lymer B L, Olsen S, Lundqvist, J and Lindström A, 2014 Water governance and management challenges in the continuum from land to the coastal sea–spatial planning as a management tool

[3] Wiering M and Immink I 2006 When water management meets spatial planning: a policy-arrangements perspective *Environment and planning C: Government and policy* **24(3)**423-38

[4] van der Kwast J, Yalew S, Dickens C, Quayle L, Reinhardt J, Liersch S, Mul M, Hamdard M and Douven W 2013 A framework for coupling land use and hydrological modelling for management of ecosystem services *Int J. Env. Mon. and Anal.* **1(5)**230-6

[5] Arciniegas G and Janssen R 2012 Spatial decision support for collaborative land use planning workshops *Lands. Urban Plan.* **107(3)**332-42

[6] Schilling K E, Jha M K, Zhang Y K, Gassman P W and Wolter C F 2008 Impact of land use and land cover change on the water balance of a large agricultural watershed: Historical effects and future directions *Water Resour. Res.* **44(7)**

[7] Ghaffari G, Keesstra S, Ghodousi J and Ahmadi H 2010 SWAT-simulated hydrological impact of land use change in the Zanjanrood basin, Northwest Iran *Hydro. Proc: An Int. J.* **24**(7)892-903

[8] Neitsch S L, Arnold J G, Kiniry J E A, Srinivasan R and Williams J R 2002 Soil and water assessment tool user's manual version 2000 *GSWRL report* **202(02-06)**

[9] Karvonen T, Koivusalo H, Jauhiainen M, Palko J and Weppling K, 1999 A hydrological model for predicting runoff from different land use areas *J Hydrol.* **217(3-4)**253-65

[10] Naef F, Scherrer S and Weiler M 2002 A process based assessment of the potential to reduce flood runoff by land use change *J Hydrol.* **267(1-2)**74-9

[11] Senseman GM, Bagley C F and Tweddale S A 1995 *Accuracy Assessment of the Discrete Classification of Remotely-Sensed Digital Data for Landcover Mapping* (No. CERL-TR-EN-95/04) CONSTRUCTION ENGINEERING RESEARCH LAB (ARMY) CHAMPAIGN IL

[12] Landis J R and Koch G G 1977 The measurement of observer agreement for categorical data. *Biometrics* 159-174

[13] Koeln G and Bissonnette J 2000 Cross-correlation analysis: mapping landcover change with a historic landcover database and a recent, single-date multispectral image In *Proc. 2000 ASPRS Annual Convention Washington DC*

[14] Lunetta R S and Elvidge C D 1999 *Remote sensing change detection* **(Vol. 310)** Taylor & Francis

[15] Lillesand T M, Kiefer R W, Dulbahri, Suharsono P, Hartono, Suharyadi and Sutanto 1993 *Penginderaan Jauh dan Interpretasi Citra* Gadjah Mada University

[16] Gassman P W, Williams J R, Benson V R, Izaurralde R C, Hauck L M, Jones C A, Atwood J D, Kiniry J R and Flowers J D 2005 Historical development and applications of the EPIC and APEX models *2004 ASAE Annual Meeting* American Society of Agricultural and Biological Engineers

[17] Neitsch S L, Arnold J G, Kiniry J R and Williams J R 2011 *Soil and water assessment tool theoretical documentation version 2009* Texas Water Resources Institute

[18] Smarzyńska K and Miatkowski Z 2016 Calibration and validation of SWAT model for estimating water balance and nitrogen losses in a small agricultural watershed in central Poland *J Water Land Dev.* **29**(1) 31-47

[19] Nash J E, Sutcliffe J 1970 River flow forecasting through conceptual models part I—A discussion of principles *J Hydrol.* **10** 282–290

[20] Eisenhauer J G 2003 *Regression through the origin* Teaching Statistics **(25)**76–80

[21] Moriasi D N, Arnold J G, Van Liew M W, Bingner R L, Harmel R D and Veith T L 2007 Model evaluation guidelines for systematic quantification of accuracy in watershed simulations *Transactions of the ASABE* **50(3)** 885-900

[22] Ridwansyah I, Pawitan H, Sinukaban N and Hidayat Y 2014 Watershed modeling with

[23] ArcSWAT and SUFI2 in Cisadane catchment area: Calibration and validation of river flow Prediction *J Adv Sci Eng Inf Technol* **6(2)**92-101

[24] Yini H, Jianzhi N, Zhongbao X, Wei Z, Tielin Z, Xilin W, and Yousong Z 2016 Optimization of land use pattern reduces surface runoff and sediment loss in a hilly-gully watershed at the Loess Plateau, China *Forest Syst.* **25**(1)9

[25] Wang H, Sun F, Xia J, and Liu W 2017 Impact of LUCC on streamflow based on the SWAT model over the Wei River basin on the Loess Plateau in China *Hydrol. Earth Syst. Sci.* **21**(4) 1929

[26] Baker T J and Miller S N 2013 Using the Soil and Water Assessment Tool (SWAT) to assess land use impact on water resources in an East African watershed *J Hydrol.* **486** 100-111

[27] Marhaento H, Booij M J, Rientjes T H M and Hoekstra A Y 2017 Attribution of changes in the water balance of a tropical catchment to land use change using the SWAT model *Hydrol. Process* **31**(11) 2029-40

[28] Githui F, Mutua F and Bauwens W 2009 Estimating the impacts of land-cover change on

runoff using the soil and water assessment tool (SWAT): case study of Nzoia catchment, Kenya *Hydrolog. Sci. J* **54**(5)899-908
[29] Bracmort K S, Arabi M, Frankenberger J R, Engel B A and Arnold J G 2006 Modeling long-term water quality impact of structural BMPs *Trans. ASABE* **49**(2) 367-74
[30] Betrie G D, Mohamed Y A, Griensven A V and Srinivasan R 2011 Sediment management modelling in the Blue Nile Basin using SWAT model *Hydrol. Earth Syst. Sci.* **15**(3) 807-818
[31] Xie H, Chen L and Shen Z, 2015 Assessment of agricultural best management practices using models: Current issues and future perspectives *Water* 7(3)1088-1108
[32] Taylor N 2010 Commentary: What is this thing called spatial planning? an analysis of the British government's view *Town Planning Review* **81**(2) 193-208
[33] Hersperger A M, Oliveira E, Pagliarin S, Palka G, Verburg P, Bolliger J and Grădinaru S 2018 Urban land-use change: The role of strategic spatial planning *Global Environ. Chang.* **51** 32-42
[34] Memarian H, Balasundram S K, Abbaspour K C, Talib J B, Boon Sung C T and Sood A M 2014 SWAT-based hydrological modelling of tropical land-use scenarios *Hydrolog. Sci. J.* **59**(10) 1808-29
[35] J G, 2007 Spatial planning, water and the Water Framework Directive: insights from theory and practice *Geogr. J.* **173**(4) 330-42
[36] Guswa A J, Brauman K A, Brown C, Hamel P, Keeler B L and Sayre S S 2014 Ecosystem services: Challenges and opportunities for hydrologic modeling to support decision making *Water Resour. Res.* **50**(5) 4535-44
[37] Maidment D R, Hooper R P, Tarboton D G and Zaslaksky I 2009 Accessing and sharing data using CUAHSI water data services In Hydroinformatics in hydrology, hydrogeology and water resources Proc.Symposium JS. 4 at the Joint Convention of the International Association of Hydrological Sciences (IAHS) and the International Association of Hydrogeologists (IAH) held in Hyderabad, India, 6-12 September 2009 213-223 IAHS Press

Acknowledgments
This study supported by Priority Research Program of Deputy of Earth Science, the Indonesian Institute of Sciences (LIPI), entitled **"Response of water ecosystem to climate change and anthropogenic as the basis for the development of coastal region development strategies"**, funded by LIPI through the fiscal year of 2015-2017.

Assessing phytoplankton distribution and water quality in constructed wetlands during dry and wet periods: A Case Study in USM Engineering Campus

S Shaharuddin[1*], NA Zakaria, Ab Ghani and Wan Maznah WO[2]

[1] River Engineering and Urban Drainage Research Centre (REDAC), Universiti Sains Malaysia, Seri Ampangan, 14300 USM, Nibong Tebal, Penang, Malaysia
[2]School of Biological Sciences, Universiti Sains Malaysia, 11800 USM, Penang, Malaysia

**Corresponding author: redacsyafiq@usm.my*

Abstract. Constructed wetlands (CW) are built to improve water quality while serving as an alternative sustainable habitat for aquatic life. In Malaysia, CWs are designed according to the guideline for Urban Stormwater Management Manual for Malaysia (MSMA 2nd Edition) 2012. This study shall evaluate the performance of constructed wetland design as stipulated by MSMA through assessment of water quality and phytoplankton growth, which acts as a biodiversity indicator. A 5-month sampling period with phytoplankton abundance in a constructed wetland was correlated with the selected water quality index (WQI) parameters comprising dissolved oxygen (DO), ammoniacal oxygen (AN), chemical oxygen demand (COD), biological oxygen demand (BOD), total suspended solid (TSS), and pH. The constructed wetland consists of three main zones, namely the forebay, macrophytes and micropool zones. The highest WQI value was obtained from the micropool zone with a range of 78.98-85.45 (82.71±3.35) while the lowest WQI value was collected from the macrophytes zone with a range of 68.13-77.96 (73.25±3.42). A total of 15 phytoplankton species from 5 different algal phyla were identified in the constructed wetland. Phytoplankton distribution was high in the macrophytes zone (521 cell m^{-3}) during dry season due to high nutrient concentration as opposed to the other zones. The phytoplankton distribution was found to be closely associated with DO, AN, TSS and total phosphorus. At the same time, the macrophyte zone design also affects the distribution of phytoplankton.

1. Introduction

Phytoplankton not only serves as a primary producer in the food chain but also plays a significant part in the primary productivity system. In addition, it acts as a bio-indicator for numerous environmental factors aside from being a biological filter by accumulating and removing nutrients in the aquatic ecosystem. The diversity of algal or phytoplankton community is used in assessing the level of anthropogenic stress to the ecosystem [1]. Thus, it is necessary to measure environmental impacts of phytoplankton community due to land use changes such as the introduction of stormwater constructed wetlands. Constructed wetlands are engineered systems that have been designed and built to utilize the natural processes involving wetland vegetation, soils, and their associated microbial assemblages to assist in treating pollution and wastewater [2]. The free water surface (FWS) constructed wetland (CW) is a popular type of constructed wetland to treat wastewater, as indicated by [3]. Among the types of wastewater treated by FWS CW are residential stormwater runoff [4] and highway stormwater runoff [5]. A typical FWS CW with emergent macrophytes constitutes a shallow sealed basin or sequence of basins, containing 20-30 cm of rooting soil, with the water depth of 20-40 cm. FWS CWs are effective in removing organics through microbial degradation and settling of colloidal particles [6]. Treatment studies of FWS CWs have also been well documented in tropical climate regions, especially in Malaysia [7-11].

The Urban Stormwater Management Manual for Malaysia (MSMA) is a standard design guideline of the Department of Irrigation and Drainage (DID) Malaysia [12]. Most evaluation and assessments pertaining to water quality studies are based on the first edition of MSMA, implemented since the year 2000. In 2012, the second edition of MSMA was introduced with various improvements in regard to design, monitoring, etc. Being new, only a few assessments and studies have been conducted to evaluate freshwater phytoplankton growth in reference to MSMA Second Edition. Thus, this study hopes to evaluate the distribution of freshwater phytoplankton in a constructed wetland and correlate such distribution with water quality as an environmental factor. Here, water quality shall be evaluated in accordance with the water quality index (WQI) as proposed by the Department of Environment (DOE) Malaysia.

2. Materials and methods

2.1. Site description

The constructed wetland in this study is located in Universiti Sains Malaysia (USM) Engineering Campus (Figure 1). Its exact coordinates are between latitudes 100°29.5' South and 100°30.3 North and between longitudes 5° 9.4' East and 5° 8.5' West in Mukim 9 of Seberang Perai Selatan District, Pulau Pinang. The locality is known as Sri Ampangan, Nibong Tebal, Pulau Pinang which is about 2 km south-east of Nibong Tebal town centre, 1.5 km north-east of Parit Buntar (Perak) and nearly 1.5 km north-west of Bandar Baharu (across Kerian River in Kedah). The campus area is about 320 acres in size and made up of mainly flat land used for oil palm plantation. This constructed wetland receives storm water runoff from 17.6 acres catchment area of faculty buildings and car park areas.

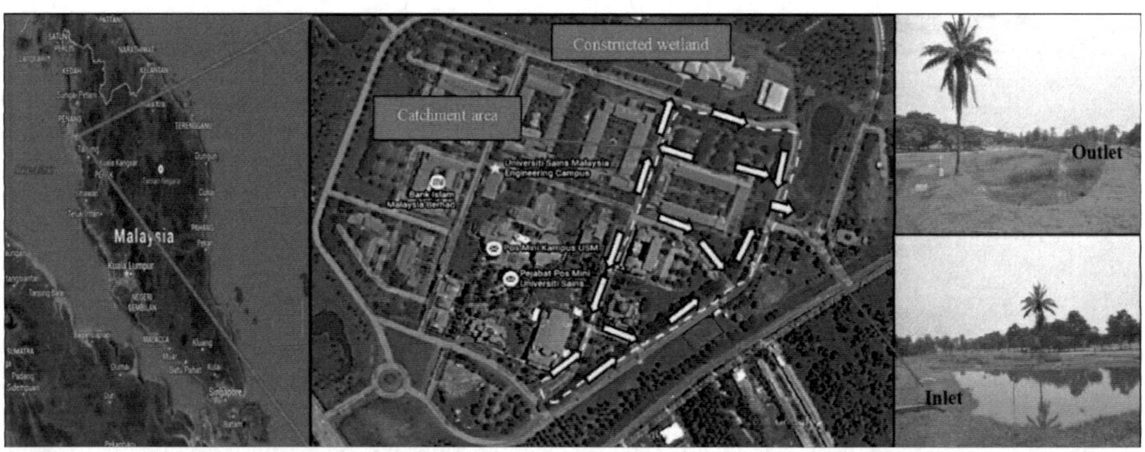

Figure 1. Location of study area of stormwater constructed wetland in USM Engineering Campus.

2.2. Sampling and analysis

The constructed wetland comprises three zones, namely the forebay, macrophyte and micropool zones. The forebay is the initial point to receive discharge from drainage/swale; while the macrophyte zone is composed of high-marsh, low-marsh and deep-marsh zones. This zone is teeming with macrophytes or wetland plant species especially from the emergent types as explained in detail in Table 1. The last zone before water exits the constructed wetland is called the micropool. Table 1 also shows the water depth measurement and types of emergent plants present. Sampling was conducted in a once-a-month basis throughout the period beginning November 2014 until March 2015. Four (4) sampling points were selected in the forebay (IF, FA, FB and FC), seven (7) points were chosen in the macrophytes zone (IAM, IBM, MA, MB, MC, OAM and OBM) and three (3) sampling point were selected in the micropool zone (MCA, MCB, OM). In addition, the study employed a 3-hour sampling duration from 9 am until 12 pm to collect grab samples, with a standard 0.5m below surface water. The phytoplankton were sampled using plankton net, filtered 40L samples were preserved using formalin 5% solution and Lugol concentration. Identification was performed using the taxonomic keys [13] and [14]. Enumeration of phytoplankton samples was carried out in line with the method recommended [15]. The in-situ parameters were determined at the sampling point during the sampling process using YSI Professional Plus handheld multi-parameter meter while laboratory analysis was conducted in accordance with HACH, Water Analysis Handbook 5th Edition. Further, precipitation data were collected using rain gauge placed in the vicinity of the constructed wetland, while Pearson correlation and regression was utilized to quantify the relationship between phytoplankton abundance (cell/ml) and WQI.

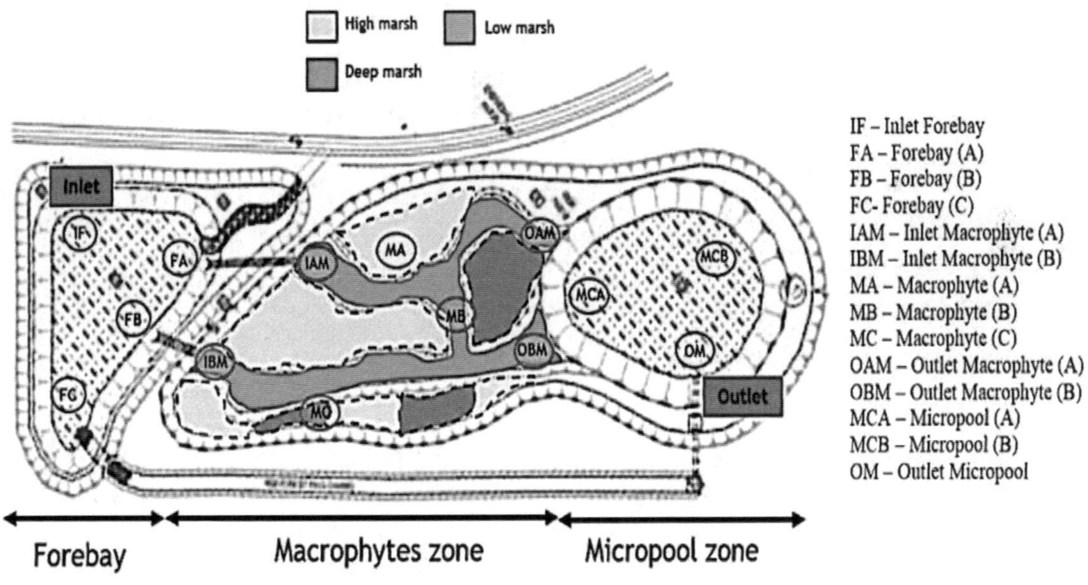

Figure 2. The sampling point and zonation of constructed wetland; Forebay, Macrophytes zone (High marsh, Low marsh and Deep marsh) and Micropool zone.

Table 1. Zone characteristics (water depth and plants/macrophytes) in the constructed wetland.

Zonation		Water depth (m) (min-max)	Types of plants	Plant Density	Picture
Forebay		0.6-1.0	-	-	-
Macrophyte Zone	**High Marsh**	0-0.3	-*Donax grandis*, -*Eleocharis variegata*	1.63 ind/m^2	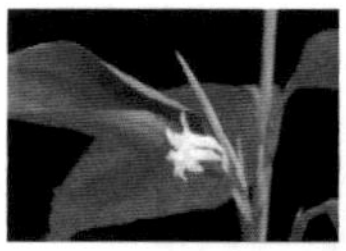 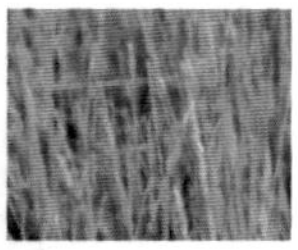
	Low Marsh	0.3-0.5	*Phragmates karka*	58 ind/m^2	
	Deep Marsh	0.6-1.0	*Typha angustifolia*	29 ind/m^2	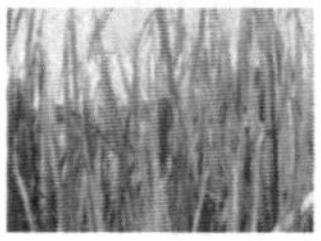
Micropool		0.7-1.4	-	-	-

In this study, the water quality index (WQI) from the Department of Environment (DOE) Malaysia has been adopted to measure water quality based on the given formula;

$$WQI = (0.22 * SIDO) + (0.19 * SIBOD) + (0.16 * SICOD) + (0.15 * SIAN) + (0.16 * SISS) + (0.12 * SIpH)$$

Where;

SIDO = Sub-index DO (% saturation); SIBOD = Sub-index BOD; SICOD = Sub-index COD

SIAN = Sub-index NH3-N; SISS = Sub-index SS/TSS; SIpH = Sub-index pH

Whereas the sub-index was calculated based on the following formula;

Sub-index for DO (In % saturation)

$SIDO = 0$ for $x \leq 8$

$SIDO = 100$ for $x \leq 92$

$SIDO = -0.395 + 0.030x^2 - 0.00020x^3$ for $8 < x < 92$

Sub-index for BOD

$SIDOD = 100.4 - 4.23x$ for $x \leq 5$

$SIDOD = 108 * \exp(-0.055x) - 0.1x$ for $x > 5$

Sub-index for COD

$SICOD = -1.33x + 99.1$ for $x \leq 20$

$SICOD = 103 * \exp(-0.0157x) - 0.04x$ for $x > 20$

Sub-index for NH3-N

$SIAN = 100.5 - 105x$ for $x \leq 0.3$

$SIAN = 94 * \exp(-0.573x) - 5 * I\ x - 2\ I$ for $0.3 < x < 4$

$SIAN = 0$ for $x \geq 4$

Sub-index for SS/TSS

$SISS = 97.5 * \exp(-0.00676x) + 0.05x$ for $x \leq 100$

$SISS = 71 * \exp(-0.0061x) + 0.015x$ for $100 < x < 1000$

$SISS = 0$ for $x \geq 1000$

Sub-index for pH

$SIpH = 17.02 - 17.2x + 5.02x^2$ for $x < 5.5$

$SIpH = -242 + 95.5x - 6.67x^2$ for $5.5 \leq x < 7$

$SIpH = -181 + 82.4x - 6.05x^2$ for $7 \leq x < 8.75$

$SIpH = 536 - 77.0x + 2.76x^2$ for $x \geq 8.75$

Note:

* Means multiply with

3. Results and discussion

3.1. Monthly precipitation and water temperature

Based on rainfall measurement as shown in Figure 3, the months of November and December 2014 were classified as wet season, with the highest rainfall amount measuring 17.5 mm in November and 27.7 mm in December. The graph also shows November to experience higher rainfall frequency compared to December. Dry season, on the other hand, occurred from January 2015 until the end of the sampling period in March 2015. Rainfall had influenced water temperature in the constructed wetland as thermometer reading was below 30°C, while the lowest reading at 28.09°C was obtained in December 2014. Water temperature rose during dry season, as thermometer reading peaked at 30.99°C in February. This is probably due to water in the constructed wetland being exposed to the sun for a longer period, which subsequently caused gradual temperature to rise throughout the sampling period. These three main factors, namely rainfall amount, water temperature and duration of sun exposure, will impact the concentration/value of water quality parameters as well as phytoplankton distribution and abundance.

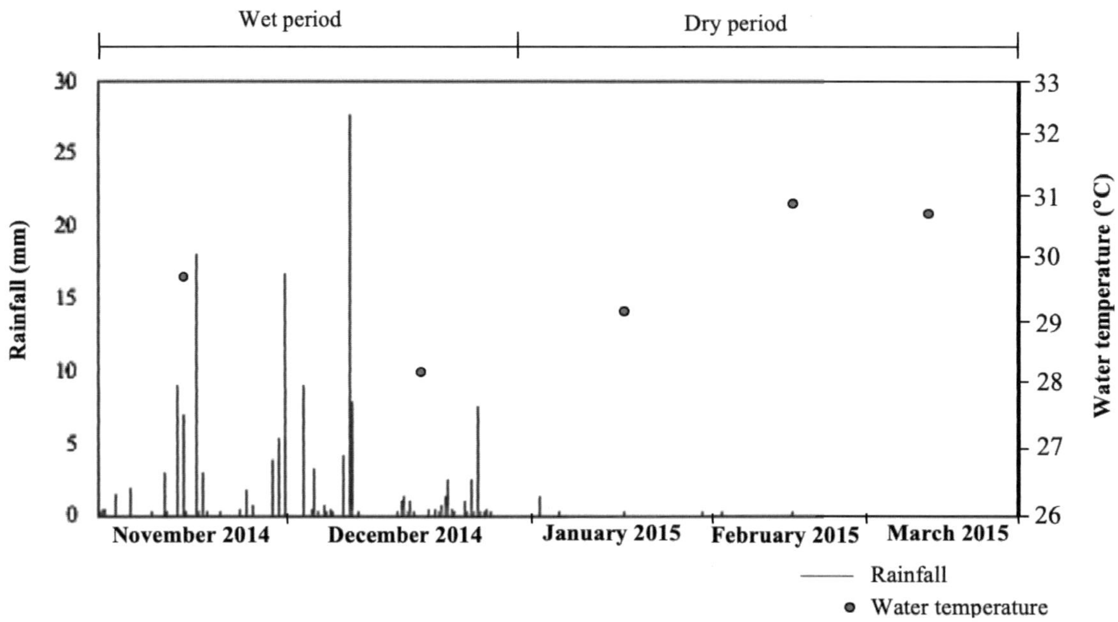

Figure 3. Rainfall (mm) pattern and water temperature (°C) from November 2014 until March 2015.

3.2 Spatial and temporal water quality concentration

Figure 4 shows the monthly variation in water quality as well as the different sampling point results throughout the sampling period. These six parameters make up the main parameter for evaluation as they represent WQI value, which shall sum up the water quality between November 2014 and March 2015. Based on the results, ammoniacal nitrogen (AN) showed significant variation in terms of monthly comparison, whereby low AN value was obtained during the wet season as opposed to dry season. The abundance of water from runoff and precipitation had diluted AN concentration in the months of November and December 2014 in the constructed wetland. From the box plot, AN concentration decreased from the inlet Forebay to the outlet Micropool. The micropool area produced Class II AN in which the lowest concentration of 0.11 mg/l was obtained at OM. For the other parameters, there were variations between the forebay, macrophytes and micropool zones. For BOD5, a high concentration range of 2.12-2.37 mg/l was obtained at the Forebay, while low concentration range between 1.40-2.07 mg/l was obtained at the macrophytes zone. Biodegradable organic matter from microorganism and nutrient concentration affected BOD concentration. In other cases, oxygen production from the phytoplankton may have contributed to the amount of oxygen in the water, which affected degradation process from microorganism as well BOD reading. As for the range of COD concentration, most sampling points fell into Class III, with 24 mg/l and 21 mg/l except for MC and OM. Low COD concentration of 7 mg/l was obtained at MC whereas OM had 9 mg/l. This also shows that organic matter was reduced, starting from the inlet at the forebay before it exits the micropool. The level of organic matter can be ascertained by referring to the TSS concentration. TSS concentration experienced a reduction starting from the inlet at the forebay before it leaves the OM, micropool. In addition, organic particles from decomposing materials might also contribute to TSS concentration. As algae, plants and animals decay, the decomposition process allows small organic particles to break away and enter the water column as suspended solids. However, most sampling points indicate Class I TSS concentration, with reference to the box plot graph. The pH value indicates acceptable ranges between 6.5-8.5, with December 2014 showing low pH value while February 2015 showing high pH value. The pH value indicates low pollutant level in the constructed wetland and this argument can be supported by looking at all the graph plots. High DO concentration at the Forebay compared the other zones may be due to sufficient nutrients for phytoplankton to undergo photosynthesis with less amount of microorganism for degradation process. Compared to the macrophyte zone, although the nutrient amount was high, the amount of microorganism is predicted to be high as well due to the high number of plant density. The root plants offer a place for bacteria and microorganism to grow, which use oxygen for the degradation process [16]. Figure 5 shows the different nutrient concentration mean of total nitrogen (TN) and total phosphorus (TP) at each sampling point, with the highest mean obtained at the macrophytes zone, at 0.19-0.34 mg/l for TP and 2.35-16.47 mg/l for TN. The high nutrient concentration obtained in the macrophytes zone was due to the nutrient uptake capability of plants. Jinadasa and team [17] had found that for satisfactory plant establishment (which in this case in the newly planted plant in the newly constructed macrophytes zone) certain period of time required for their establishment such as *Typha* sp took 7-8 month for their establishment for nutrient uptake as well as growth. Plant decay also contributed to the high nutrient in the zone. However, such a problem can be solved through monitoring of the abundant plants and maintaining each marsh area via annual harvesting. This factor had also affected phytoplankton distribution as the density was high at the macrophytes zone compared to the other zones due to nutrient availability.

3.3. Monthly and sampling point variation of phytoplankton abundance

Figure 6 presents the phylum of phytoplankton distribution throughout the sampling period. Four main phylum were observed with Chlorophyta being the most abundant phylum at 585-770 cell/ml, followed by Bacillariophyta (239-298 cell/ml), Cyanophyta (59-103 cell/ml), Chrysophyta (21-46 cell/ml) and Pyrrhophyta (9-24 cell/ml). Chlorophyta and Bacillariophyta were the most commonly sampled phylum and are major inhabitants of the freshwater as well as other aquatic ecosystems [18,19]. In addition, Figure 7 shows some of the species observed and obtained in the constructed wetland. The species of phytoplankton in the wetland is an important food source for herbivorous species [20]. Olurin and Awolesi [21] documented that the *Tilapia mariae* and *Chromidotilapia guntheri* fish species, mainly feed on phytoplankton consisting of the desmid genera *Closterium* and *Cosmarium*. Thus, *Closterium* sp suits this constructed wetland well due to the presence of the *Tilapia* sp. fish.

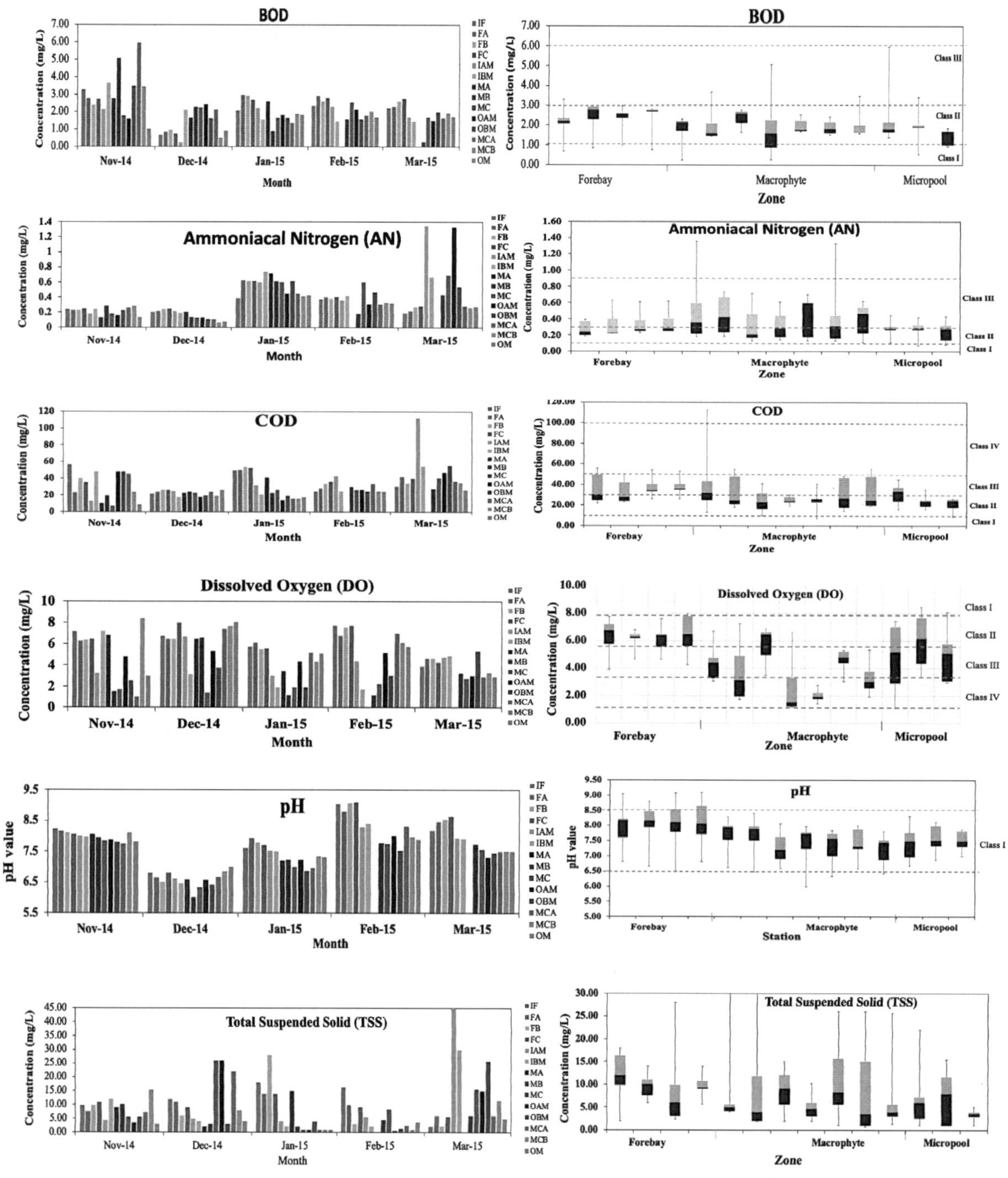

Figure 4. Monthly variation (left side) and sampling point variation (right) of mean concentration of six parameters; (from top to bottom) BOD, Ammoniacal nitrogen (AN), COD, DO, pH, and TSS from November 2014 to March 2015.

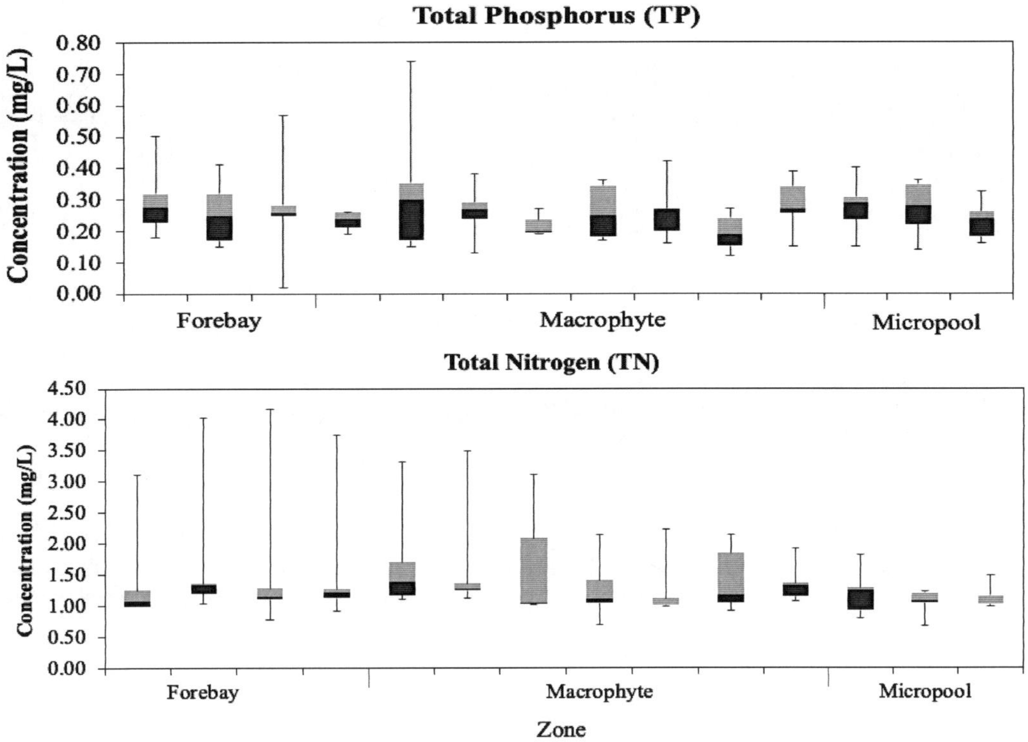

Figure 5. Mean total phosphorus (TP) and total nitrogen (TN) concentration from November 2014 until December 2015 in the sampling point constructed wetland

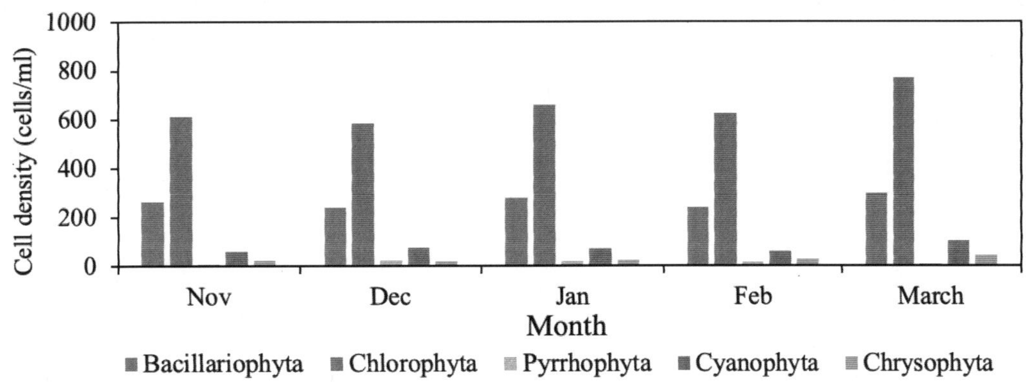

Figure 6. Monthly distribution of phytoplankton density by phylum in the constructed wetland from November 2014 until March 2015.

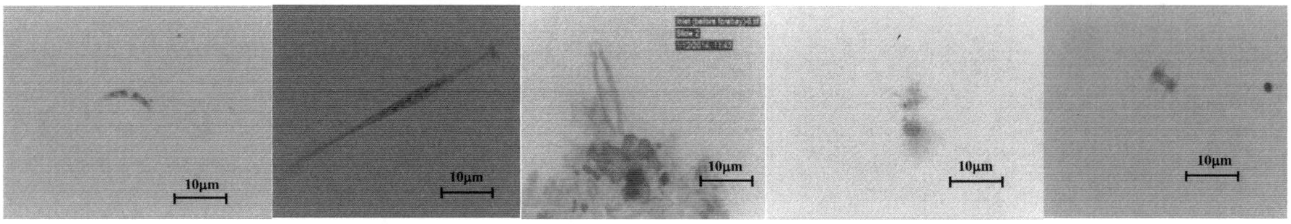

Figure 7. Freshwater phytoplankton species (From left: *Closterium parvulum, Nitzschia longissimi, Stauroneis anceps., Lyngbya confervoides, Staurastrum sp.*)

3.4. Relationship between phytoplankton, WQI and water quality parameters

Based on Pearson correlation information in table 2, DO concentration showed a significant high negative relationship with up to 0.01 significance level for phytoplankton phylum, except for Phyrrophyta. The pH value indicated a positive relationship only for Chrysophyta ($p<0.05$). For the other four parameters in WQI, no significance result was found for phytoplankton density. Phytoplankton density affected DO concentration in the water. Phytoplankton decomposed by bacteria was in the constructed wetland. Bauerfeind [22] reported that bacteria play an important role in using oxygen to breakdown the phytoplankton carbon during the process of sedimentation. Thus, the energy released as organic matter is returned to the main food chain through a microbial loop of bacteria and protozoa [23]. As for the nutrient, only TN showed a significant positive relationship ($p<0.05$) with Cyanophyta and Chrysophyta. TN typically stimulates phytoplankton growth in the freshwater ecosystem in lakes and wetlands [5]. However, TP registered no significant result for all phytoplankton and this outcome could be due to source of nutrient.

Table 2. Correlation between phytoplankton and water quality parameter.

Parameters	Pearson correlation coefficient				
	Bacillariophyta	Chlorophyta	Pyrrhophyta	Cyanophyta	Chrysophyta
BOD	-0.309	-0.253	-0.089	-0.29	-0.263
AN	0.336	0.339	0.196	0.091	0.053
COD	-0.185	-0.201	-0.052	-0.406	-0.414
DO	**-0.723****	**-0.836****	-0.526	**-0.734****	**-0.836****
pH	0.430	0.477	0.115	0.380	**580***
TSS	0.236	0.144	0.336	-0.006	-0.005
TN	0.502	0.428	0.435	**0.629***	**0.623***
TP	-0.131	-0.097	0.133	-0.188	-0.117
Temp	**-0.794****	**-0.847****	**-0.729****	**-0.699****	**-0.721****
WQI	**-0.720****	**-0.852****	**-0.593***	**-0.650***	**-0.736****

** Statistically significant at the 0.01 level ; * Statistically significant at the 0.05 level

Based on the information in Figure 8, WQI value developed from the six main parameters showed all a significant negative correlation ($p<0.05$) with all phytoplankton groups. As WQI value decreased in the macrophytes zone to 71.6-77.9, phytoplankton density increased which may be due to the amount of nutrients in this zone as compared to the rest. There are several possibilities which may contribute to this zone having low WQI value compared to the others. First, it may be due to the high amount of nutrient in this area, as indicated by the graph of AN in Figure 4 as well as TN and TP in Figure 5. The source of nutrient may originate from natural processes involving decomposition of plants and animals as well as surface water runoff entering this zone from the forebay. The WQI value also showed suitability of the macrophytes zone for the phytoplankton to inhabit as well as characteristics of the area itself, which had low water depth and was likely to have received direct sunlight for photosynthesis to occur. The outlet WQI value of this constructed wetland was high, with the range value of 83.7-85.5 which showed that the constructed wetland had improved the water quality received from the catchment area, treated and polished it, before it was discharged. Johari [24] showed the reduction of nutrient from inlet to outlet of constructed wetland had improve the WQI value to be high, thus it will affect the density of phytoplankton as nutrients play a vital role in the distribution and growth of phytoplankton [25].

In Table 2, it can be seen that DO and water temperature were the most important parameters affected by the abundance of phytoplankton. Figure 9 showed how such relationships were formed in the simple regression analysis. The relationship in DO is complex as it involves a lot of factors including the respiration and decomposition rate of bacteria, which affects the oxygen concentration level in the water. Bacteria is attached to the plants parts. The more plants, the more likely places for the bacteria to inhabit and it is presumed to have high decomposition rate, as the roots and rhizomes in the soil provide surface for bacteria and other microorganisms attachment [7].Thus, the graph showed that as the density of phytoplankton is reduced, the concentration of dissolved oxygen increases. Smith and Piedrahita [26] found that high algal biomass has only an indirect linkage to low dissolved oxygen. Algal density affects turbidity, inorganic nutrient levels, and algal respiration. These, in turn, affect oxygen levels by influencing net primary production and total plankton respiration. The positive physiological effects of temperature increases might eventually turn into negative effects as soon as the optimal temperature range is exceeded [27]. This can be caused by cellular stress, due to high membrane fluidity and enhanced degradation rates of proteins and enzymes, and if these effects become significant they can negatively affect primary production [28]

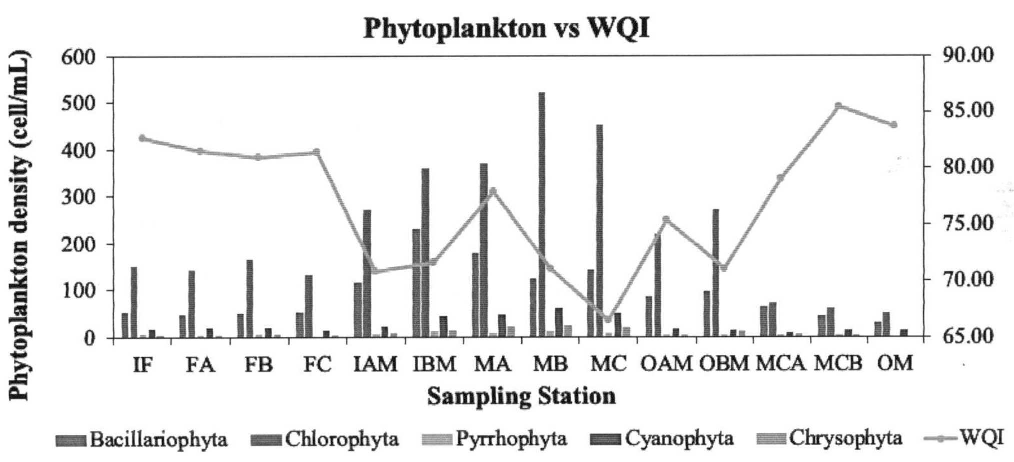

Figure 8. WQI and phytoplankton density pattern for sampling station in the constructed wetlands

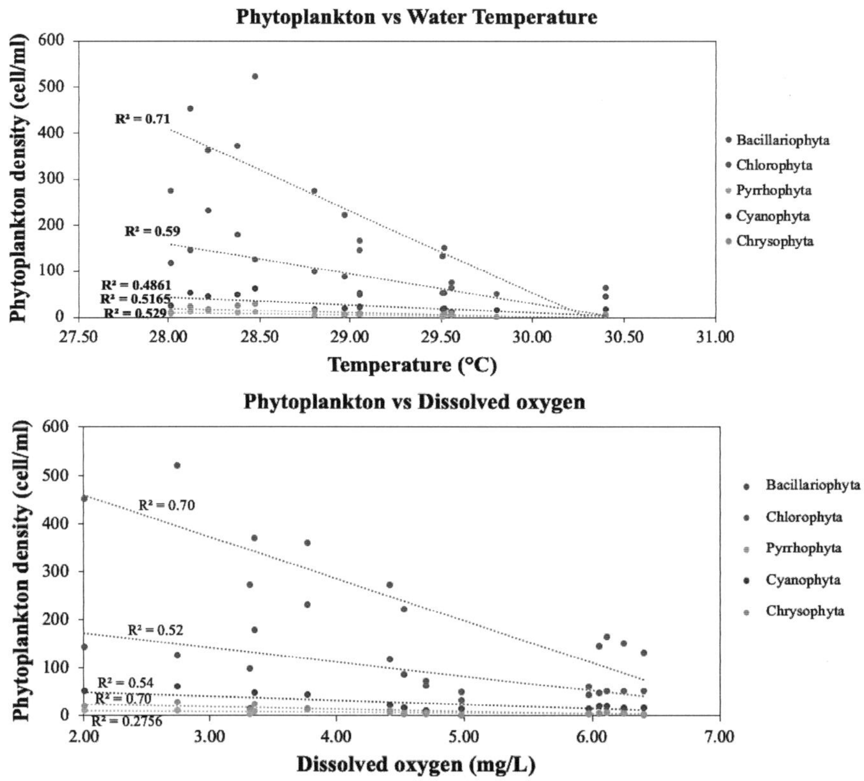

Figure 9. Relationship using simple regression analysis between temperature and dissolved oxygen (DO) towards phytoplankton density.

4. Conclusion

As a conclusion, this study proves that WQI has negative correlation with the density of phytoplankton. From the six parameters which constituted the WQI value, only DO and pH showed significant correlation with the distribution of phytoplankton density. Only TN showed slight relationship with some phylum of phytoplankton while water temperature showed significant impacts in relation to all phylum of phytoplankton. There was a slight difference in the monthly distribution of phytoplankton, as it slightly increased in density from the wet season to the dry season. However, by comparing the results from the sampling points, the distribution of phytoplankton was found to be high in the macrophytes zone compared to the other zones in the constructed wetland. As for the WQI parameters, only AN indicated an obvious difference between wet and dry seasons, while the rest only showed differences between the sampling points. For the future study, in this constructed wetland must take into consideration as they may affect the whole calculation, including WQI parameters and phytoplankton density.

5. References

[1] Lowe R.L. and Pan Y 1996 Benthic algal communities as biological monitors *In: Algal ecology freshwater benthic ecosystem Stevenson RJ, Bothwell ML, and Lowe RL, editors. New York: Academic Press* pp 753

[2] Vymazal J 2007 Removal of nutrient in various type of constructed wetland *Science of the Total Environment* **380** pp 48-65

[3] Vymazal J 2010 "Constructed Wetland for Wastewater Treatment-Review" *Water* **2** pp 530-549

[4] Bavor HJ, Davies CM and Sakadevan K 2001 Stormwater treatment: do constructed wetlands yield improved pollutant management performance over a detention pond system? *Water Science Technology* **44** pp.565-570

[5] LV J, Hongjuan W and Mengqiu C 2011 Effects of nitrogen and phosphorus on phytoplankton composition and biomass in 15 subtropical, urban shallow lakes in Wuhan, China *Limnologica* **41** pp 48-56

[6] Mohd. Noor NA, Ayub KR, Mohd Sidek L, Zakaria NA, Ab. Ghani A and Abdullah R 2004 Stormwater Treatment Using Constructed Wetland *1st International Conference on Managing Rivers in the 21st Century: Issues & Challenges, 21st - 23rd September, Penang, Malaysia* pp. 442-446

[7] Mohammadpour R, Shaharuddin S, Chang CK, Zakaria NA and Ab Ghani A 2014 Spatial Pattern Analysis for Water Quality in Free Surface Constructed Wetland *Water Science and Technology* **70**(7) pp 1161-1167

[8] Mohammadpour R Shaharuddin S Chang C K Zakaria N A Ab Ghani A Chan N W 2015 Prediction of Water Quality Index in Constructed Wetlands using Support Vector Machine *Environmental Science and Pollution Research* **22**(8) pp 6208-6219

[9] Mohd Noor NA, Mohd Sidek L, Zainal Abidin MR and Ab Ghani A 2014 Removal Efficiency Evaluation for Constructed Wetland in Tropical Climate *Journal of Energy and Environment* Universiti Tenaga Nasional pp 1-6

[10] Shaharuddin S, Chan NW, Zakaria NA, Ab Ghani A, Chang CK and Roy R 2014 Constructed Wetlands as a Natural Resource for Water Quality Improvement in Malaysia *Natural Resources* **5** pp 292-298

[11] Sim C H Yusoff M K Shutes B Ho S C Mansor M 2008 Nutrient removal in a pilot and full scale constructed wetland, Putrajaya city, Malaysia *Journal Environmental Management* **88** pp 307–317

[12] Department of Drainage and Irrigation (DID) 2012 Urban Stormwater Management Manual for Malaysia. Kuala Lumpur: Department of Irrigation and Drainage (DID) Malaysia.

[13] Tomas CR 1997 Identifying Marine Phytoplankton, Academic Press, An Imprint of Elsevier. pp.591-715

[14] Reynolds CS 1984 Phytoplankton periodicity: the interactions of form, function and environmental variability *Freshwater Biology* **14**(2) pp 111-142

[15] Leupold MA 1988 Water quality assessment. In C.S. Lobban, D.J. Chapman & B.P. Kremer (Eds.) Experimental Phycology: A laboratory Manual, Cambridge Australia pp.47-55

[16] Qomariah S Ramelan AH Sobriyah Setyno P 2016 Use of macrophyte plants, sand & gravel materials in constructed wetlands for greywater treatment *IOP Conf. Series: Materials Science and Engineering* ***176*** *012018*

[17]. Jinadasa KBSN Tanaka N Sasikala S Werellagama DRIB Mowjood MIM Ng WJ 2008 Impact of harvesting on constructed wetlands performance – a comparison between *Scirpus grossus* and *Typha angustifolia Journal of Environmental Science and Health Part A* **43**(6) pp.1-8

[18] Graham L E Graham J M Wilcox L W 2008 Algae *2nd edition. Benjamin Cummings (Pearson), San Francisco, CA*

[19] Leliaert F, Smith DR, Moreau H, Herron MD, Verbruggen H, Delwiche CF and De Clerck O 2012 Phylogeny and molecular evolution of the green algae *Critical Reviews in Plant Sciences* **31** pp 1–46.

[20] Coesel PFM 1997 The edibility of Staurastrum chaetoceras and Cosmarium abbreviatum (desmidiaceae) for Daphnia galeata/hyalina and the role of desmids in the aquatic food web *Aquatic Ecology* **31** pp 73–78

[21] Olurin KB and Awolesi O O 1991 Food of some fishes of Owa Stream, south-western Nigeria *Archiv für Hydrobiologie* **122** pp 95–104

[22] Bauerfeind S 1985 Degradation of phytoplankton detritus by bacteria: estimation of bacterial consumption and respiration in an oxygen chamber *Marine Ecology – Progress Series* **21** pp.27-36
[23] Markert BA, Breure AM and Zechmeister HG 2003 Bioindicator and Biomonitors: Principle, Concepts and Applications. Trace Metal and Other Contaminants in the Environment *Elsevier Science Publisher* **6** pp 25.
[24] Johari NE Chang CK Ab. Ghani A Ab. Wahid M Abdul-Talib S 2016 Performance of Constructed Wetland on Stormwater Quality Control IAHR APD 2016, 28-31 August, Colombo, Sri Lanka
[25] Ault T, Velzeboer R and Zammit R 2000 Influence of nutrient availability on phytoplankton growth and community structure in the Port Adelaide River, Australia: bioassay assessment of potential nutrient limitation *Hydrobiologi* **429** pp.89–103.
[26] Smith DW and Piedrahita RH 1988 The relation between phytoplankton and dissolved oxygen in fish pond *Aquaculture* **68** pp.249-265
[27] Striebel M Schabhüttl S Hodapp D Hingsamer P Hillebrand H 2016 Phytoplankton responses to temperature increases are constrained by abiotic conditions and community composition *Oecologia* **182** pp.815–827
[28] Staehr PA and Sand-Jensen K 2000 Seasonal changes in temperature and nutrient control of photosynthesis, respiration and growth of natural phytoplankton communities *Freshwater Biology* **51** pp.249–262.

Acknowledgements
The authors would like to acknowledge the technical staff from River Engineering and Urban Drainage Research Centre (REDAC) for their effort and cooperation throughout the full length of the study. Our gratitude is also extended for the financial assistance from the Ministry of Education (MOE) Malaysia in recognition of Higher Centre of Excellence (HICOE) in services to River Engineering and Urban Drainage Research Centre (REDAC), Universiti Sains Malaysia (USM). This study was funded under the Short Term Grant Research USM (Grant Number: 304/PREDAC/60313014). Credit also goes out to REDAC's Director, Prof Nor Azazi Zakaria, my main supervisor, Prof Aminuddin Ab. Ghani and my co-supervisor Prof Dr Wan Maznah Wan Omar for their support throughout this research. Many thanks to the School of Biological Science for organizing the International Conference on Ecology and Biodiversity Across Time and Space (ICEBATS) as well.

Comparative limnology of natural and man-made tropical lakes

Z Sharip[1*], FM Yusoff[2], J Jusoh[3] and A Jamin[4]

[1] Lake Research Unit, Water Quality and Environment Research Centre, National Hydraulic Research Institute of Malaysia, Seri Kembangan, Malaysia
[2] Department of Aquaculture, Faculty of Agriculture/International Institute of Aquaculture and Aquatic Sciences, Universiti Putra Malaysia, Serdang, Malaysia
[3] Consultating Water Expert, Ipoh, Perak
[4] PERHILITAN Tasik Bera Ramsar Site, Bera, Malaysia

**Corresponding author: ztsharip1@gmail.com*

Abstract. This paper assessed and compared physicochemical profiles and biodiversity indices of two shallow lakes namely the natural Bera Lake and Bukit Merah reservoir in Peninsular Malaysia. Variations in relation to hydro-meteorological variables, including rainfall, wind speed and air temperature were also measured during the field sampling. The results showed clear distinctions between natural and man-made lakes in terms of physical and chemical properties of lake waters and their relationships to the hydro-meteorological dynamic. Persistent oxygen and temperature differences between the epilimnion and hypolimnion were clearly detected in Bukit Merah reservoir under low winds and strong heat. Oxycline also persist during flooding or high turbidity levels indicating their enriched bottom waters. On the other hand, the natural Bera lake showed temporary stratification during the dry season with low winds but the water columns were thoroughly mixed during the wet season when the water temperature was lower and wind speed was higher. Plankton diversity in Bera lake was significantly higher ($p < 0.05$) than the Bukit Merah reservoir, probably due to the much longer history of the natural lake compared to the man-made one. Due to the longer residence time of the reservoir compared to the natural lake, the former tends to be more vulnerable to eutrophication if subjected to uncontrolled nutrient enrichment.

1. Introduction

Extensive limnological studies in Malaysia have been undertaken since 1970s [1] starting with Bera lake followed by Paya Bungor and Ulu Lepar, Kenyir and Loagan Bunut [2,3]. To date various limnological studies including lake profiles and lessons learned briefs have been published in approximately 26 lakes in the country such as the Chini lake, Temengor reservoir, Putrajaya lake, Batang Ai reservoir and Loagan Bunut lake [4]. Most of these studies were focused on one lake and based on field experiment alone. In-depth comparative studies of lakes were limited to provide generalization for management purposes. Understanding such limnological characteristics is crucial for management and governance improvement.

In this study, we characterize the limnological features of natural and man-made shallow lakes in Malaysia and their seasonal physico-chemical-biological changes using field experiments and numerical simulation results. The main objective of the study was to illustrate the limnological characteristics of natural and man-made lakes in order to provide better understanding of their variation to aid in developing strategies for sustainable management of lakes by combining results from field measurements and modelling. Bera lake and Bukit Merah reservoir, representing natural and man-made lake respectively, were selected due to the reported catchment pressures that affects their water quality and trophic state [5, 6]. Bera lake is the oldest and largest freshwater lake in Malaysia while Bukit Merah reservoir is one of the oldest reservoirs created for irrigation and flood mitigation [5, 6]. These two lakes were shallow and are important sites for ecological biodiversity and conservation.

2. Materials and method

Sampling were performed at both Bera Lake and Bukit Merah reservoir (figure 1) during a wet season (October - December) and dry season (March - May) at three sites. *In situ* measurements of physico-chemical profiles (depth, temperature, dissolved oxygen, pH and turbidity) were carried out at approximately 10 cm below the lake surface till the deepest depth using a multi-parameter probe, YSI 6600 (YSI Inc., Yellow Springs. OH, USA). Water samples were collected using a Van Dorn sampler and was preserved in cooler boxes at 4 °C and transported to laboratory for analysis of chlorophyll-a (Chl *a*) using fluorometric method and total phosphorus (TP) using persulphate digestion and automated ascorbic acid reduction method [7]. Water for chlorophyll-a samples was filtered on site to prevent pigment deterioration.

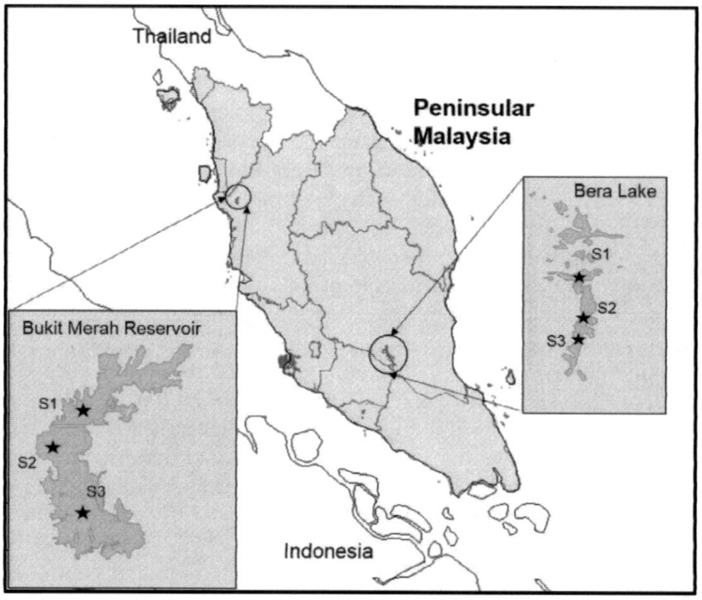

Figure 1. Study site and sampling area.

For phytoplankton, one litres of water samples were collected below the surface of water column of the lake and samples were fixed with Lugol's iodine to achieve 1% final solution and transported to the laboratory. In the laboratory, all preserved samples were left to settle to the bottom of the measuring cylinder. The samples were concentrated to a 100 ml working volume and identified to the lowest taxa based on descriptions, taxonomic keys and illustrations in [8, 9] and densities were computed in terms of number of cells/ml. For zooplankton, samples were collected in vertical hauls using a 63 μm plankton net by lowering the net to near lake bottom to the surface. The samples were poured into 500 ml bottles and fixed with buffered formalin to provide 5% formalin final solution. Samples were collected in duplicates at each site for each season. The zooplankton samples were identified to the lowest taxa [10, 11] and densities were counted as the number of individuals per litre of water sample.

Daily meteorological data were collected at two meteorological stations that were deployed at the two sites for measurement of wind speed, rainfall and water temperature (figure 2). Additional physical chemical data under different meteorological conditions were obtained from simulation results of a well-calibrated model that have been established using 3-dimensional model using Coastal Environment Simulation System (CENSIS) software [12]. Simulation were carried out under calm condition (wind speed <0.5 m/s), average wind condition and maximum wind speed (>7 m/s).

Figure 2. Weather monitoring stations deployed for the study.

All environmental variables and plankton data were assessed for normality prior to statistical testing. Nutrient, phytoplankton and zooplankton data were log-transformed to improve normality and balance contribution of the overall abundances. Species diversity and richness were examined using the Shannon-Wiener diversity index (H') and Margalef species richness index (d), respectively. Independent t-test was used to detect differences in environmental variables between the two lakes. All statistical analyses were performed using SPSS version 22.

3. Results and discussion

3.1. Physico-chemical profiles

The results showed clear distinctions between natural and man-made lakes based on mean and ranges of environmental variables (table 1). Due to its surrounding forested catchment, the minimum water temperature in Bera lake was lower than Bukit Merah reservoir ($p < 0.05$). Bukit Merah reservoir showed higher mean DO and pH levels. Low dissolved oxygen and pH levels in Bera Lake indicated its dystrophic environment. Higher dissolved organic contents and humic acids have been reported in swampy lakes such as Bera, indicating the presence of peaty environmental condition and characteristics of a brown/black water system [6, 13]. Chlorophyll-a values were also low likely due to low pH levels that limits phytoplankton growth [13]. Trophic state index based on chlorophyll-a indicated Bera lake is oligotrophic-mesotrophic while Bukit Merah reservoir is mesotrophic-eutrophic. Previous study has illustrated that the trophic status of Bukit Merah reservoir has fluctuated between mesotrophic and eutrophic states [6, 14].

Table 1. Mean ± SD and ranges (in brackets) of physico-chemical parameters between Bera and Bukit Merah lakes.

	Bera lake	Bukit Merah reservoir	*p* value
Temperature (°C)	29.7 ± 0.9 (28.1 – 31.2)	30.3 ± 0.85 (29.4 – 32.2)	0.047
Dissolved oxygen (mg/L)	4.6 ± 1.8 (2.8 – 7.3)	7.2 ± 0.8 (4.8 – 8.3)	0.000
pH	5.6 ± 0.6 (4.2 - 5.8)	6.7 ± 0.7 (5.8 – 8.3)	0.000
Turbidity (NTU)	8.2 ± 6.6 (1.6 – 21.7)	23.2 ± 32.8 (1.2 – 146.3)	0.02
Total phosphorus (mg/L)	0.04 ± 0.04 (<0.01 – 0.16)	0.04 ± 0.03 (<0.01 – 0.12)	0.676
Chlorophyll-a (µg/L)	4.0 ± 4.1 (<0.5 – 11.1)	9.5 ± 8.9 (<0.5 – 36.7)	0.005
Trophic State Index (Chl *a*)	23.8 – 57.1	30.6 – 65.9	

3.1.1. Dry season.

Dry season was observed during the south-west monsoon season that occurs from May to September. In the natural Bera lake, warmer temperature was observed in deeper sites with temperature differences of 3 ºC between the surface and 1.5 m depth (figure 3a). In the man-made Bukit Merah reservoir, vertical temperature pattern varied between sites. Generally surface temperature was cooler than the surface temperature in Bera lake indicating the influence of wind speed (average wind speed was >1.0 m/s). Thermal layer at sites deeper than 3 m were not weakened by the wind probably due high density differences between water layers, a typical condition for a tropical lake. Both temperature and turbidity differences with depth might have contributed to the persistent thermocline. Persistent oxygen differences between the epilimnion and hypolimnion were detected in both lakes under low winds and strong heat. Dissolved oxygen levels were high (> 6 mg/l) near the surface to about 2 m depth and DO concentration started to decrease at deeper depths in both lakes (figure 3c, d). High wind events provided good mixing and increased the surface mixed layer in the large man-made lake.

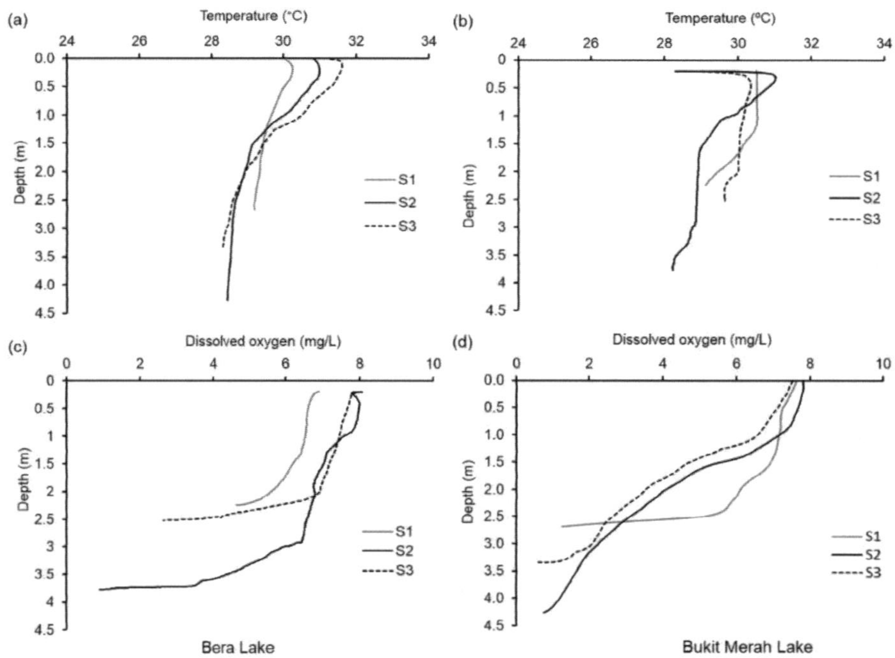

Figure 3. Vertical profiles of temperature and dissolved oxygen during the dry season.

3.1.2. Wet season.

Wet days were observed during the north-east monsoon season which normally occurs from November to March. Water levels in the natural Bera lake almost doubled during the wet season over the study period with depth variations of 4.5 m during the low water (dry days) to 9.0 m in the high-water period (wet days). Surface temperature dropped from 31.2 °C during dry days to 28.1°C during wet days. Similarly, surface DO levels in Bera Lake dropped to below 6 mg/L during the wet days. Oxycline persisted in these wet days possibly due to flooding or high turbidity levels from the backflow of the Pahang river indicating their enriched bottom waters. Bera lake also showed weak thermal stratification during the wet season with low winds and the water columns were thoroughly mixed when the water temperature was lower and wind speed was higher (figure 4a). Physico-chemical profiles in Bera Lake were consistent with other floodplain lakes located in the black water system in the Amazon, Brazil [15].

IOP Conf. Series: Earth and Environmental Science **380** (2019) 012019 doi:10.1088/1755-1315/380/1/012019

In contrast, the man-made Bukit Merah lake did not experience significant increase in water levels during this wet period due to water level regulation. Vertical temperature profiles highly varied with sites. Thermal bar was formed at 1.0 m depth for sites with water columns deeper than 2 m. Oxycline also persisted during flooding or high turbidity levels indicating their enriched bottom waters. Dissolved oxygen levels remained high (> 6 mg/L) in Bukit Merah reservoir from the surface to a depth of about 2 to 3 m (figure 4d). Diurnal changes on thermal regime are well reported in tropical lakes located in Brazil with diurnal stratification during the day and nocturnal mixing during the night which shape vertical variation of other chemical properties such as pH and chlorophyll-a [16,17]. Thermal structure induced by solar heating are weakened by stronger wind action [12, 18]. Numerous studies have shown the influence of meteorological conditions on the hydrodynamic of lakes, which are in turn influenced by the latitude [18, 19]. Wind speed and its direction do not only shape thermal mixing and surface mixed layer in Shikinawa Lake but also an important transport mechanism of pollutants in Putrajaya lake [18, 20]. Latitudinal gradient played an important role in solar irradiance which differentiate the limnological characteristics in tropical lakes with the ones in the temperate zone [21].

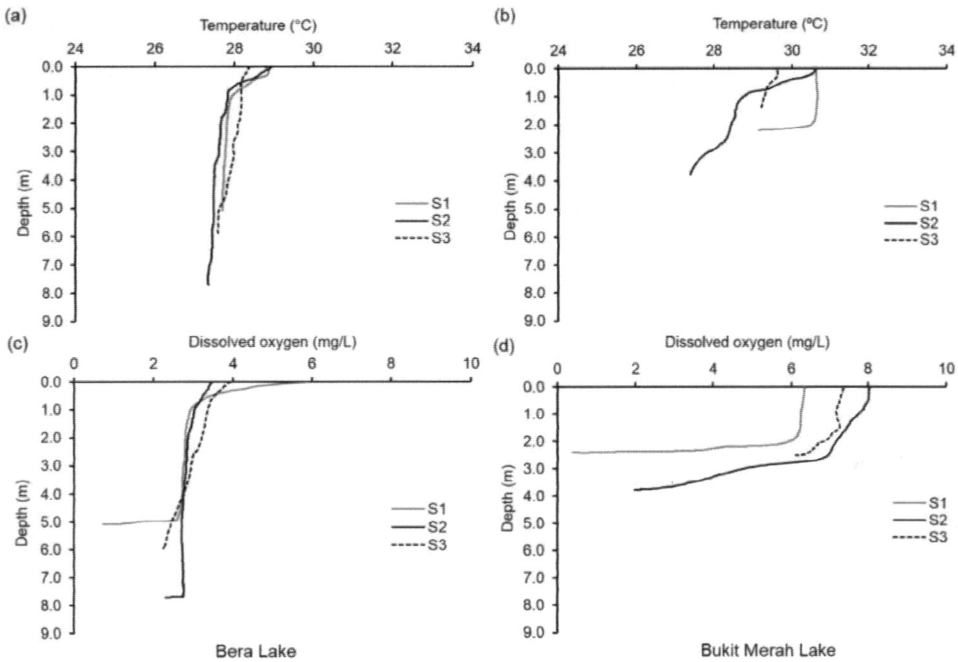

Figure 4. Vertical profiles of temperature and dissolved oxygen during wet season.

3.1.3. Calm and windy conditions.

Wind speed varied seasonally at the two lakes (figure 5a). Calm condition usually occurred in the early morning during the sampling period. Highest hourly wind speed was recorded in Bukit Merah reservoir during the wet season (~4 m/s, around 4 pm) while the highest hourly wind speed was recorded in Bera lake during the dry season (~ 3.2 m/s, around 3 am). Mean daily wind speeds were almost half in Bera lake compared to the mean daily wind speeds in Bukit Merah reservoir (figure 5b). Higher wind speeds in Bukit Merah reservoir could be higher due to larger open surface area surrounded by oil palm plantations, and thus higher wind fetch. Bera Lake is more dendritic in shape and a large part of the lake is covered by wetland marshes while its shoreline is sheltered by forest, and these terrestrial and aquatic vegetation protected the lake from wind effects [6, 13].

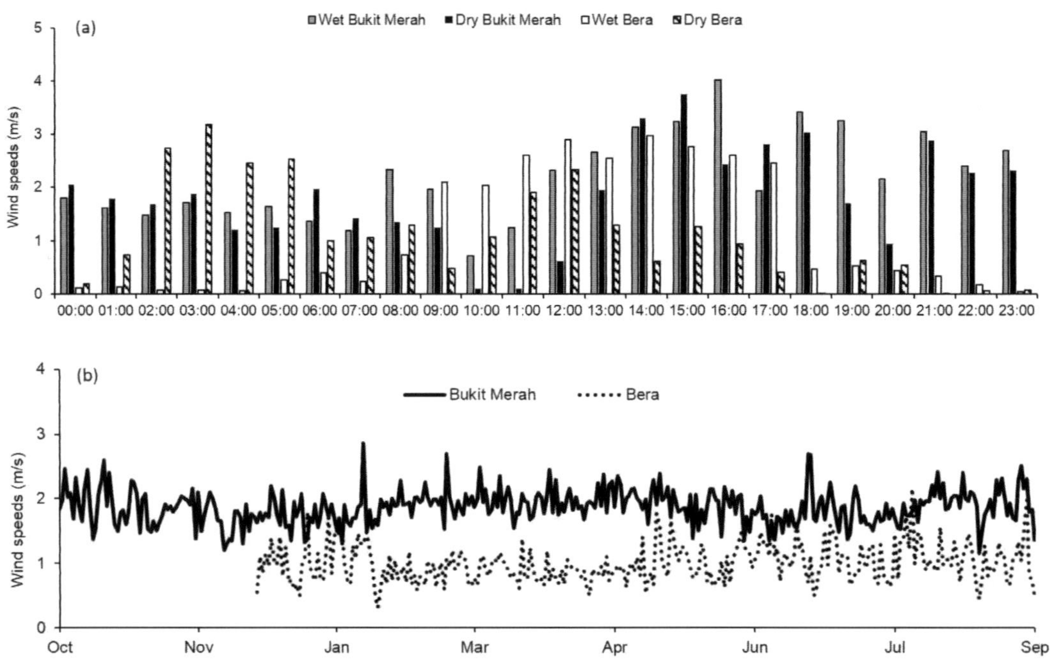

Figure 5. Wind speed profiles (a) diurnally during wet and dry seasons (b) daily from wet season (Oct) to dry season (Sept).

High wind speeds provided good mixing throughout the Bukit Merah reservoir (figure 6). High winds events resulted in not only a well-mixed temperature profile spatially, but also vertically throughout the water column. During calm winds, spatial variations of temperature were apparent; warmer temperature were localised in the sedimented area in the east while cooler temperature was observed in the southern area dominated by submerged vegetation. In Bera lake, isothermic conditions persist at all sites under calm condition during the late evening and early morning. Temperature-depth profile showed weak pattern of stratification during the day despites weak wind conditions (figure 7). Lake thermal structure and the associated nutrient availability and extent of euphotic zone influenced the vertical distribution of primary productivity in Lake Tanganyika [21].

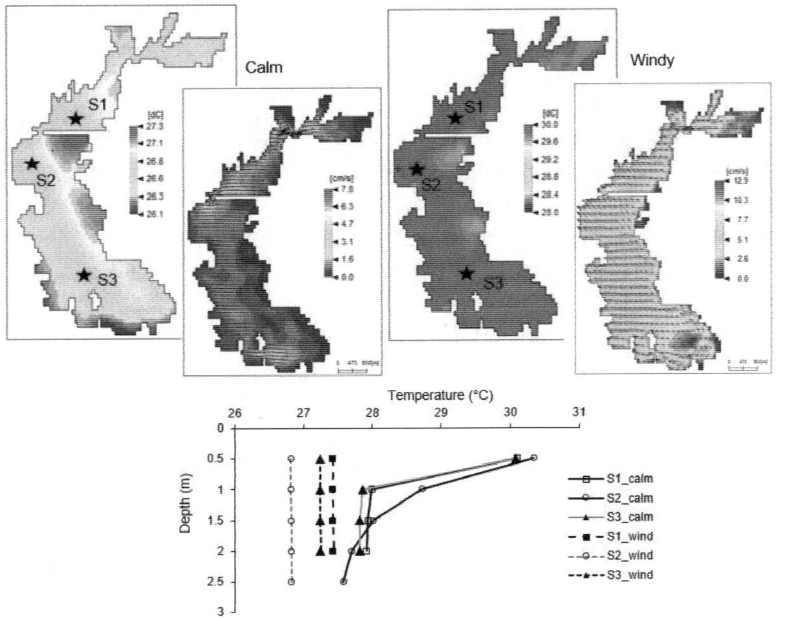

Figure 6. Spatial and vertical temperature-depth profiles during calm and windy conditions in Bukit Merah reservoir.

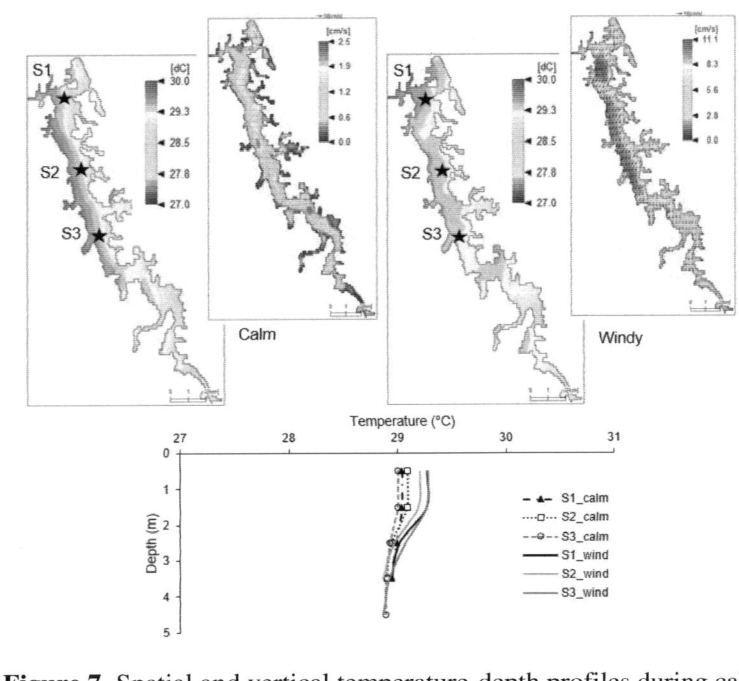

Figure 7. Spatial and vertical temperature-depth profiles during calm and windy conditions in Bera lake.

3.2. Plankton profiles

The total phytoplankton density was significantly higher in Bukit Merah reservoir compared to Bera lake ($p = 0.001$). Lower phytoplankton density could be due to lower pH levels that limit growth as well as the presence of massive macrophytes stands that limits nutrient availability for phytoplankton [9]. Species richness was significantly different ($p < 0.05$) between lakes. Higher phytoplankton species richness in Bera lake than Bukit Merah reservoir was possibly due to the longer history of the former, which was estimated to be more than 4500 years [22].

In terms of zooplankton, total density was higher in Bukit Merah reservoir compared to Bera lake ($p = 0.003$). Higher zooplankton density in this man-made lake was reported to be associated to higher density of food namely the phytoplankton [13]. However, zooplankton species richness was higher in the natural Bera lake compared to the created Bukit Merah lake (table 2).

Table 2. Phytoplankton and zooplankton density and diversity in Bera lake and Bukit Merah reservoir.

	Bera Lake		Bukit Merah reservoir	
	Dry	Wet	Dry	Wet
Phytoplankton				
Total density	248.4	342.8	9350.8	17117.2
Species richness	1.78	1.99	0.84	0.89
Diversity Index	1.51	1.69	1.21	1.11
Zooplankton				
Total density	14.6	22.1	683.0	1217.0
Species richness	2.51	4.03	0.59	0.61
Diversity Index	1.21	1.14	1.37	1.15

4. Conclusion

The results of these studies showed clear differences between natural and man-made lakes in terms of physical and chemical properties of their water columns, and their relationships to the hydro-meteorological dynamic. Limnological characteristics of natural and man-made lakes were different. Unlike the man-made lake, water level of the natural lake was strongly influenced by hydrological conditions. Large man-made lake such as Bukit Merah reservoir was shaped by meteorological conditions such as wind effects. Plankton diversity in Bera lake was significantly higher than the reservoir probably due to the much longer history of the natural lake compared to the man-made one. In general, longer residence time of a reservoir compared to a natural makes the former more vulnerable to eutrophication if subjected to uncontrolled nutrient enrichment. The use of both measurements and numerical simulations provided better understanding of lake physical-chemical characteristics and their seasonal changes. This information is valuable in the establishment of sustainable management plans of the lake ecosystem.

5. References

[1] Ho S C 1980 Status of limnological research and training in Malaysia *Limnology in developing countries* ed. B Gopal B and R G Wetzel (International Association for Limnology) pp 163-189

[2] Yusoff F M 2008 Tasik Kenyir: thermal stratification and its implications on sustainable management of the reservoir *Proceedings of Colloquium on Lakes and Reservoir Management: Status and Issues* eds A Anton, S Zakaria, F M Yusoff, L K Sim, L Jin and Z Sharip (Seri Kembangan: National Hydraulic Research Institute of Malaysia) p 89-99

[3] Sharip Z and Zakaria S 2007 Lake and reservoir in Malaysia: management and research challenges. *Proc. Taal2007: the 12th World Lake Conference* Ed. Sengupta M and Dalwani R (Jaipur, India) pp 1349-55

[4] NAHRIM 2016 *Blueprint for Lake and Reservoir Research and Development (R&D) in Malaysia*

Seri Kembangan: National Hydraulic Research Institute of Malaysia p 91
[5] Hidzrami S A 2010 Bukit Merah, Perak, Lake Brief. *Managing lakes and their basins for sustainable use in Malaysia (Lake Briefs Report Series 1)* (Kuala Lumpur: Akademi Sains Malaysia) pp 29-72
[6] PERHILITAN 2012 Lake Briefs Report: Integrated Management of Ramsar Site, Tasek Bera, Pahang- a Department of Wildlife and National Parks Peninsular Malaysia Experience *Managing lakes and their basins for sustainable use in Malaysia* (Seri Kembangan: National Hydraulic Research Institute of Malaysia) pp 1-15
[7] American Public Health Association 2012 *Standard methods for the examination of water and waste water*. Washington (DC)
[8] Pentecost A (1984) Introduction to Freshwater Algae. (England: Richmond Publishing Co. Ltd.)
[9] Wehr JD, Sheath R G and Kociolek J P (2015) *Freshwater algae of North America: ecology and classification*. Academic Press, New York, p 918
[10] Idris B A G (1983) *Freshwater zooplankton in Malaysia (Crustacea: Cladocera)* (Serdang: Universiti Putra Malaysia)
[11] Phan D D, Nguyen V K, Le T N, Dang Ngoc T and Ho T H (2015) *Identification Handbook of Freshwater Zooplankton of the Mekong River and its Tributaries* (Vientiane: Mekong River Commission)
[12] Sharip Z, Shah S A, Jamin A and Jusoh J 2018 Assessing the Hydrodynamic Pattern in Different Lakes of Malaysia *Applications in Water Systems Management and Modeling*. Ed. D. Malcangio (Croatia: IntechOpen) chapter 5 pp 71-85
[13] Sharip Z and Yusoff F M 2017 *J Env Biol.* **38** 1365
[14] Najib S A M, Ismail W R and Omar M A 2017 J Env. Geogr. **10** 1
[15] Aprile F, Darwich A J, Siqueira G W, Santos F R R and Miguéis A M B 2013 *Int. Res. J Env Sci.* **2** 58
[16] Barbosa F A R and Tundisi J G 1989 *Archiv Fur Hydrobiologie* **116** 333
[17] Henry R 1995 The thermal structure of some lakes and reservoirs in Brazil *Limnology in Brazil* eds J. G. Tundisi, C E M Bicudo and T M Tundisi (Rio de janeiro) pp 351
[18] Tuan N, Hamagami K, Mori K and Hirai Y 2009 *Paddy Water Env.* **7** 83
[19] Lewis W M 1996 Tropical lakes: how latitute makes a difference *Perspectives in Tropical Limnology* eds F. Schiemer and K. T. Boland (Amsterdams: SPB Academic Publishing bv) pp 43-64
[20] Sharip Z, Saman J M, Noordin N, Majizat A, Suratman S and Shaaban A J 2016 *Modeling Earth Env. Syst.* **2** 1
[21] Mziray P, Kimirei I A, Staehr P A, Lugomela C V, Perry W L, Trolle D, O'Reilly C M and Mgana H F 2018 *J Great Lakes Res.* https://doi.org/10.1016/j.jglr.2018.08.015
[22] Wüst R A J and Bustin R M 2001 *Int J Coal. Geol.* **46** 179

Acknowledgments

This research was supported by the Ministry of Science and Technology of Malaysia, E-science Fund Grant No. 04-03-09-SF0003.

IOP Conf. Series: Earth and Environmental Science **380** (2019) 012020 doi:10.1088/1755-1315/380/1/012020

Phytoplankton composition and the occurrence of cyanobacterial bloom in Lake Maninjau, Indonesia

Sulastri[1*], C Henny[1] and A B Santoso[1]

[1]Research Centre for Limnology, Indonesian Institute of Sciences, Jl. Raya Bogor km 46, Cibinong, Indonesia

**Corresponding author: sulastri@limnologi.lipi.go.id*

Abstract. Algae blooms have been recorded in Lake Maninjau in November 2000, October 2011 and recently on April 2018. These blooms were indicated by green scum formation on the lake surface with a very high chlorophyll-a concentration, as high as > 100 µg-L. We determined the characteristics of phytoplankton composition and abundance including environmental conditions during cyanobacterial blooms and non-cyanobacterial blooms in Lake Maninjau. During cyanobacterial blooms, phytoplankton were dominated by *Microcystis aeruginosa* with a maximum abundance of 24,320 x 10^3 individual L^{-1} (94.4 % of the total assemblage). While during the non-blooming period, cyanobacteria species were more diverse, represented by *Cylindrospermopsis raciborskii*, *Anabaea affinis*, *Aphanizomeon* sp, *Planktolyngbya* sp. and *Chroococcus* sp. Diatom (*Synedra ulna*) generally occurred in all conditions, however, desmids (green algae) disappeared during cyanobacteria blooms. It is highlighted that the occurrence of *Microcystis* blooms can be related to total phosphorous dynamics in the lake.

1. Introduction

One of the most concerning trends in freshwater systems globally is the increasing distribution of toxic cyanobacteria blooms due to eutrophication and climate change [1]. In Indonesia, the occurrence of *Microcystis* blooms have been reported in Karangkates Reservoir in 2001 and in Lake Maninjau 2000 [2, 3]. In Lake Maninjau, the occurrence of *Microcystis* bloom is more frequent, in 2011[4] and April 2018. Most reservoirs and lakes in Indonesia have been heavily impacted by anthropogenic activities and floating fish culture [5]. In Lake Maninjau, organic loads from floating cages was estimated at 111,889.84 tons with an average of 9324.98 tons/year from 2001 to 2013 [6]. With a total of 16,210 units of floating cage, 12,090 tons of fish was harvested from the cage in 2013 [7]. While the trophic state has changed gradually from mesotrophic to eutrophic, succession in phytoplankton assemblage followed [8]. Wastes from the cage contribute to the increased of total phosphorus, ortho-phosphate and nitrogen compounds in the lake [9].

Environmental impact of cyanobacteria bloom to the water would depend on the duration of the bloom, the species producing the bloom, and its cell concentrations. There is a wide range of cell concentration related to the phytoplankton composition of cyanobacteria bloom event. In a lake of Bangladesh at the peak period of phytoplankton blooms, the cell density of *Microcystis aeruginosa* was 1550 x10^3 cells mL^{-1} which comprised 96.84 % of the total phytoplankton [10]. In the Hoan Kiem Lake Vietnam the number of *Microcystis* cells ranged from 27 x 10^4 to 5 x 10^8 cells mL^{-1} [11]. Another approach associated with bloom events criteria is chlorophyll-*a* concentration per unit volume of water. Specific species such as *Microcystis* has been used to develop status criteria of cyanobacteria bloom.

For instance, the Commonwealth of Virginia established blooming criteria for *Microcystis aeruginosa* under chlorophyll-*a* level of 27.5 µg L^{-1} (27.5 $mg.m^{-3}$) and 50,000 cells. ml^{-1} due to production of potential toxin [12].

The occurrence of *Microcystis* blooms in fresh water ecosystems has been explained by a series of factors related to species capacity in regulating buoyancy, water column stability, as well as environmental factors such as the availability of light, nutrients, temperature and pH [13,10, 16]. Other studies have attributed the algal blooms occurrence to physical characteristics such as lake morphometry, water column mixing, and temperature stratification [14]. For instance, in Taechung Reservoir Korea, the major factor influencing blue green dominance was a weak monsoon which was directly linked to strong water column stability, high water temperature (>28°C), and reduced silica input. Low N:P ratios were not the determining factors in this lake. Instead, light and pH level seemed to act as secondary factors [15]. In Lo Galindo Lake, Central Chile, bloom of *Microcystis* was generated when the temperature reached its annual peak of 25 °C [16].

The status of *Microcystis* bloom in Lake Maninjau together with phytoplankton assemblage, abundance and chlorophyll-*a* concentration is still absent. Thus, environment factors controlling the occurrence of the bloom have been barely identified. We determined phytoplankton assemblage and environmental factors in Lake Maninjau during cyanobacteria bloom and non-bloom period to support the management strategy in preventing harmful effect of the produced toxin.

2. Materials and methods

2.1. Research location

Lake Maninjau is a tectono-volcanic lake located in West Sumatera, Indonesia, between 100° 08'53.84" E to 100° 14'02.39" E and 0° 14' 52.50" S to 0° 24' 12.17" S (figure 1) at 462 m above sea level with a surface area of 9,737.50 ha, an average depth of 105.5 m and a maximum depth of 168 m [17]. The volume of water stored in the lake, shore line and shore line development is 10.33 billion m^3, 52.7 km and 1.51 km/km^2, respectively [17]. There are two peak rainy seasons on April to May and October to November. Rainfall is relatively high through the years and dry season is not so dominant [18].

2.2. Data collection

Surface and subsurface water samples were collected at 9 study sites (figure 1) on April 2009, June 2009, April 2014, August 2015, and March 2016 (non *Microcystis* bloom period) and 6 sites on April 2018 (*Microcystis* bloom period) to determine species composition and abundance of phytoplankton and physico-chemical parameters including Secchi depth, temperature, conductivity, dissolved oxygen (DO), pH, Total Nitrogen (TN) and Total Phosphorus (TP) and Chlorophyll-*a* concentration. In this study, we used the data of phytoplankton during the period of *Microcystis* bloom from the April 2018 observation. This is because the data of phytoplankton abundance in *Microcystis* bloom in 2000 and 2011 was limited [4]. The physical description of each research location is presented in table 1.

2.3. Physical and Chemical measurements

Water temperature, pH, Dissolved Oxygen (DO), and conductivity data were collected using a water quality checker (Horiba U-10), while water transparency was examined by the measurement of the Secchi depth. We quantified physical indices of the lake, i.e. thermocline depth and Schmidt stability, on the hourly average scale. Thermocline depth (m) was defined as the greatest rate of density change with depth, arrayed from vertical temperature profile recorded by a monitoring station platform. Schmidt stability (J M^{-2}), a measure of the resistance to physical mixing, was derived from the profile and lake bathymetric data. Chlorophyll-*a* sample was collected by filtering 200 to 500 mL of water sample through a GF/F Whatman glass filter paper and preserved by adding a saturated $MgCO_3$ solution. Determination of chlorophyll-*a* sample is based on [19]. Total Nitrogen (TN) was measured using the persulphate digestion and Brucine method [19, 20]. Total Phosphorus (TP) was determined using the persulphate digestion and Ascorbic Acid method [20]. All samples were analysed at Research Centre for Limnology, Indonesian Institute of Sciences (LIPI).

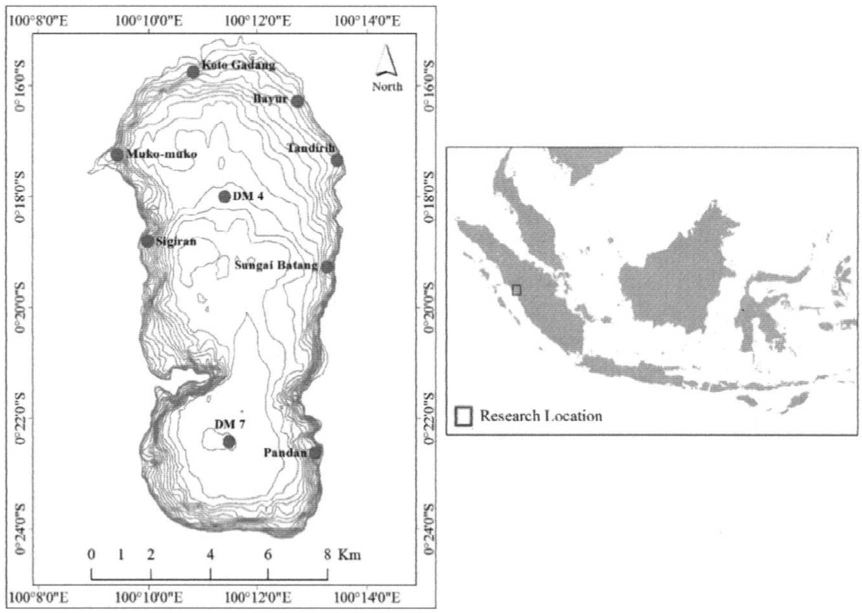

Figure 1. Lake Maninjau is located in Sumatra Island of Indonesia. The location of sampling site is shown by the blue circular. Source: GIS Lab. RC for Limnology LIPI.

Table 1. Physical description of research stations.

Name of stations and sampling period		Physical description
Apr 09, Jul 09, Apr 14, Aug 15, and Mar 16 (Non- bloom period)	Apr18 (Bloom period)	
Bayur		Area packed with cage fish culture
Tandirih	Tandirih	A few area for fish culture, the surrounding area is human settlement,
Koto Gadang		A few areas for fish culture, the surrounding area is agricultural
Muko-Muko	Muko-Muko	Near outlet (Batang Antokan), area for fish culture
Sigiran	Sigiran	Area packed with cage fish culture, the surrounding area is human settlement
Pandan	Pandan	Less area for fish culture and deep lake area
Sungai Batang	Sungai Batang	Area for fish culture, the surrounding area is agricultural
DM 4		Central part of lake (145 m depth), no fish culture
DM7	DM7	Deepest part of lake (168 m depth), no fish culture

Remark: Apr 09 =April 2009; Apr 14=April 2014; Aug 15=August 2015; Jul 09=July 2009; Mar 18=March 2018

2.4. Phytoplankton Analysis

Phytoplankton samples were collected at the euphotic zone (surface water and Secchi depth) by filtering 2 L of water through a plankton net (40 μm mesh size). The samples were preserved with 1% lugol solution for taxonomic study in the laboratory. Phytoplankton species was identified according to [21, 22, 23, 24] under magnification of 400 x through an inverted microscope. Quantitative analysis of phytoplankton used the Lackey Drop Micro Transect [25]. Direct links between environmental parameter and the percentage of dominant species of phytoplankton were analysed using Principal Component analysis (PCA) and Pearson's correlation to identify the environmental factors influencing the cyanobacteria bloom.

3. Results and Discussion

3.1. Lake stability

Lake stability is one of the environmental factors associated with a buoyancy regulating species, such as *Microcystis aeruginosa*, bloom. In Lake Maninjau, the calculated Schmidt stability shows a clear annual pattern where the lowest energy is in the early month of the year, reaching its peak in March and September (figure 2). In March to April 2018 Schmidt stability ranged from < 1000 to 4000 (J m^{-2}), and it coincided with the occurrence of *Microcystis* bloom.

3.2. Water quality

Microcystis bloom (April 2018), secchi depth transparency value ranged from 0.3 to 1.25 m (figure 3), while during the non-blooming period Secchi depth could reach 2.98 m (table 2). The highest Secchi depth transparency (2.98 m) was found in March 2016. Based on the long-term monitoring, the lowest of secchi depth transparency (< 1m) was found in the occurrence of *Microcystis* bloom as in October 2011 and April 2018 (figure 4). Hence, cyanobacteria bloom caused more light attenuation in the lake, in which other autochthonous productivity in the lake may suffer under such condition [27, 28]. There are also some important factors that influence secchi depth such as autochthonous production (plankton), cloud cover and wind action [27, 28].

Conductivity varied during non *Microcystis* bloom period and *Microcystis* bloom period (table 2). Conductivity represents ionic concentration in the waters. Higher ionic strength tended to have a higher nutrient concentration [14]. The variability of conductivity may be influenced by nutrient concentration that originated from fish culture waste. Higher pH was found in non *Microcystis* blooms period ranging from 8.44 to 8.69, while lower pH (7.75) was found in *Microcystis* blooms period, which was still in alkaline condition. It is likely that the decaying cells of *Microcystis* reduced the pH. TN and TP ratios were relatively low during the blooming period compared to the non-blooming period (table 2). Chlorophyll-*a* concentration reached 97.475 μg. L^{-1} during blooming period in April 2018 (figure 5). Based on the long-term monitoring, the highest concentration of Chlorophyll-*a* was more than 100 μg. L^{-1} (figure 6), when the occurrence *Microcystis* bloom was found in October 2011 [4].

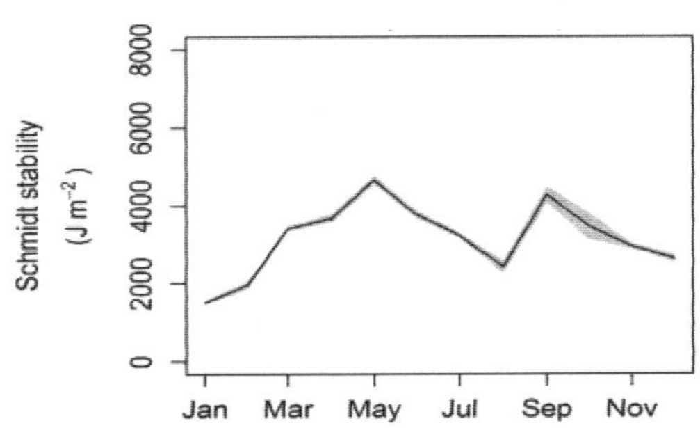

Figure 2. Annual Schmidt Stability of Lake Maninjau [26].

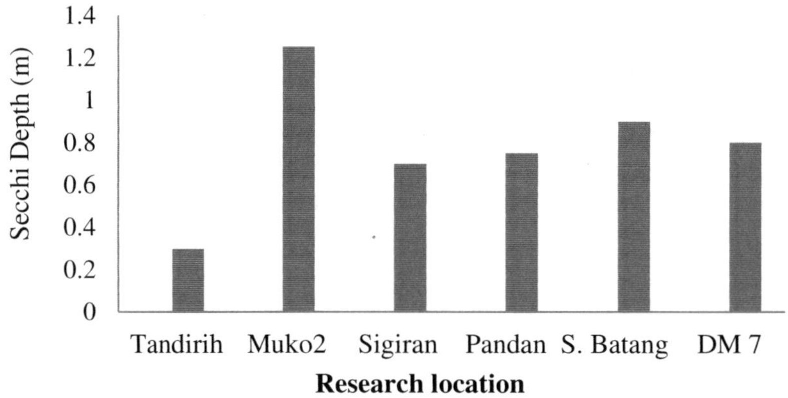

Figure 3. Secchi Depth Transparency in April 2018 (*Microcystis* bloom period).

Table 2. Average value water quality in Lake Maninjau during non *Microystis* bloom and *Microcystis* bloom period.

Parameters	Non-*Microcystis* bloom					*Microcystis* bloom
	Apr 09	Jul 09	Apr 14	Aug 15	Mar 16	Apr 18
Secchi depth (m)	2.25	1.7	2.1	1.9	2.98	0.8
Temperature (°C)	28.65	29.29	29.61	29.34	30.63	28.58
Conductivity (mS.cm^{-1})	0.113	0.100	0.132	0.129	0.110	0.130
pH	8.44	8.6	8.55	8.24	8.69	7.75
Dissolved Oxygen (DO) (mg. L^{-1})	2.88	6.00	8.21	4.35	7.76	8.71
Total phosphorus (TP) (mg. L^{-1})	0.039	0.048	0.065	0.032	0.022	0.299
Total nitrogen (TN) (mg. L^{-1})	1.486	0.582	1.058	0.974	0.415	1.117
TN:TP	39.7	13.3	21.5	48.2	20.2	4.7
Chlorophyll-*a* (µg. L^{-1})	6.21	4.55	4.00	4.80	13.67	35.31

Remark: Apr 09= April 2009; Jul 09=July 2009; Aug 15= August 2015; Mar 18=March 2018.

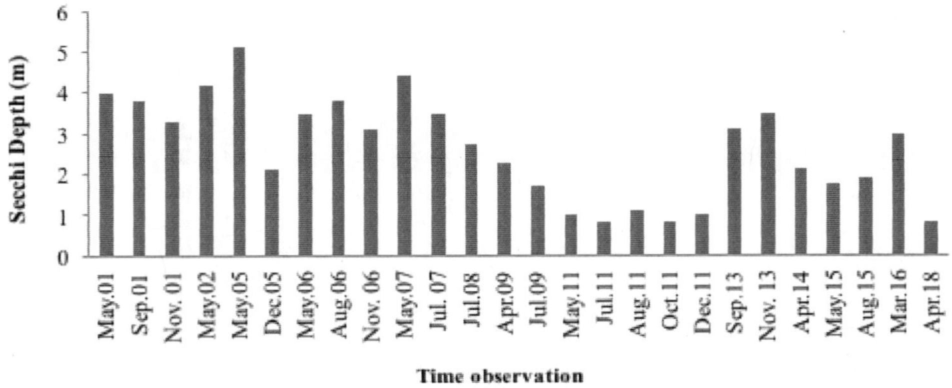

Figure 4. Long term monitoring Secchi depth transparency in Lake Maninjau (Source: Research Centre for Limnology, Indonesian Institute of Sciences Data Base).

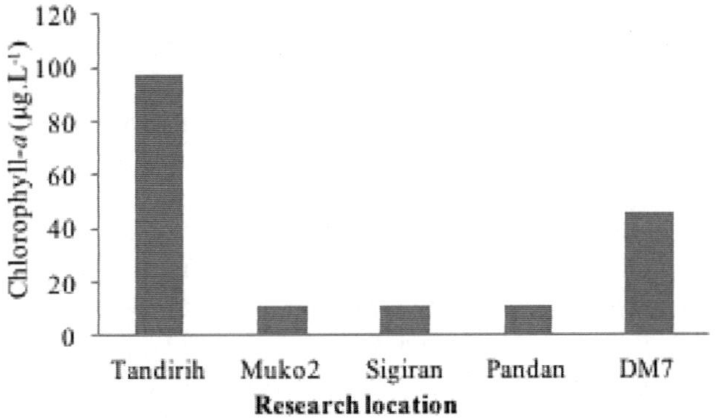

Figure 5. Chlorophyll-*a* concentration in April 2018 (*Microcystis* bloom period).

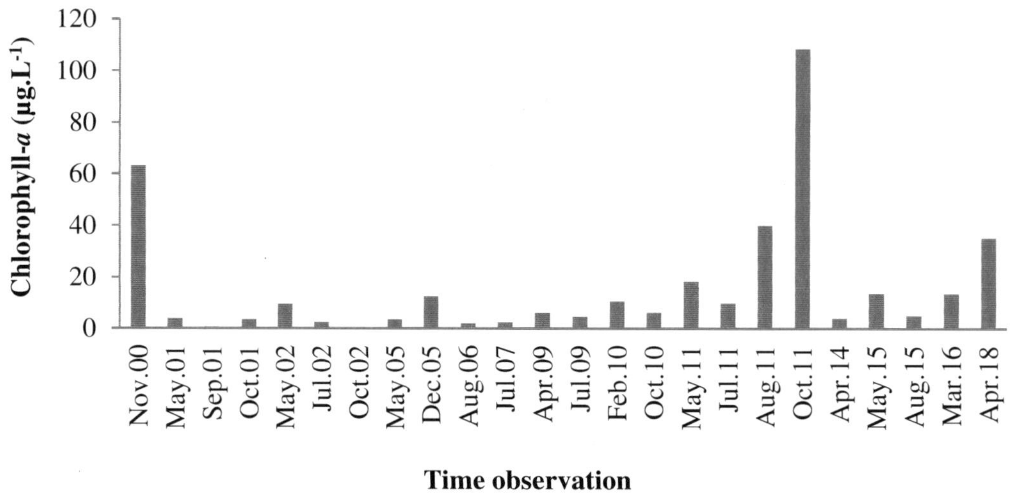

Figure 6. Long term monitoring Chlorophyll-*a* concentration in Lake Maninjau (Source: Research Centre for Limnology, Indonesian Institute of Sciences Data Base).

3.3. Phytoplankton composition and abundance

The most dominant assemblage in phytoplankton group in Lake Maninjau is Cyanophyta (blue green algae) (figure 7), except in August 2015 which was dominated by Chrysophyta. Pyrrhophyta also increased, especially *Ceratium hirrudinella.* As reported *C. hirudinella* was abundant during low phosphate and high TN:TP ratio (42.7) [29]. Our study showed that TN:TP ratio in August 2015 was 48.2 (table 2, table 3). The dominant species found in lake Maninjau was *Microcystis aeruginosa, Chroococcus* sp., *Planktoyngbya* sp., *Aphanizomenon* sp., *Anabaena affinis* and *Cylindrospermopsis raciborskii, Monallantus* sp. and *Synedra ulna* (figure 8).

The cyanobacteria species found in Lake Maninjau were *Microcystis aeruginosa, Chroococcus* sp., *Planktoyngbya* sp., *Aphanizomenon* sp., *Anabaena affinis* and *Cylindrospermopsis raciborskii* (figure 9). Most of those species are identified as toxin producer species [1, 30]. During the bloom in April 2018 *Microcystis aeruginosa* was dominant, accounted as much as 94.4 % of total assemblage. In non-cyanobacteria bloom period (March 2016), *Chroococcus sp* was the dominant species, which accounted to 86.4 % of the total composition (figure 8).

The abundance of *Microcystis* bloom period in April 2018 ranged from 37.9 to 24,320 x10^3 individual L^{-1} (table 3). This dense number of *Microcystis* cells has also been observed in the other place reaching 1.3 x 10^6 cell. L^{-1} [16]. *Anabaena affinis* and *Cylidrospermopsis raciborskii* were also abundant during *Microcystis* bloom period. These two species are commonly found in Lake Maninjau [31]. *Chrooccocus* sp. was abundant during the non-bloom period, ranging from 20.6 to 77.67 x 10^3 individual. L^{-1} (table 3). It is likely that *Chrooccocus* prefers environmental condition with high pH, and temperature and relatively better light penetration (table 4). *Synedra ulna* (diatom) always occurred in Lake Maninjau in every condition. This species is commonly found in eutrophic lake rich in nutrients and alkaline conditions [32].

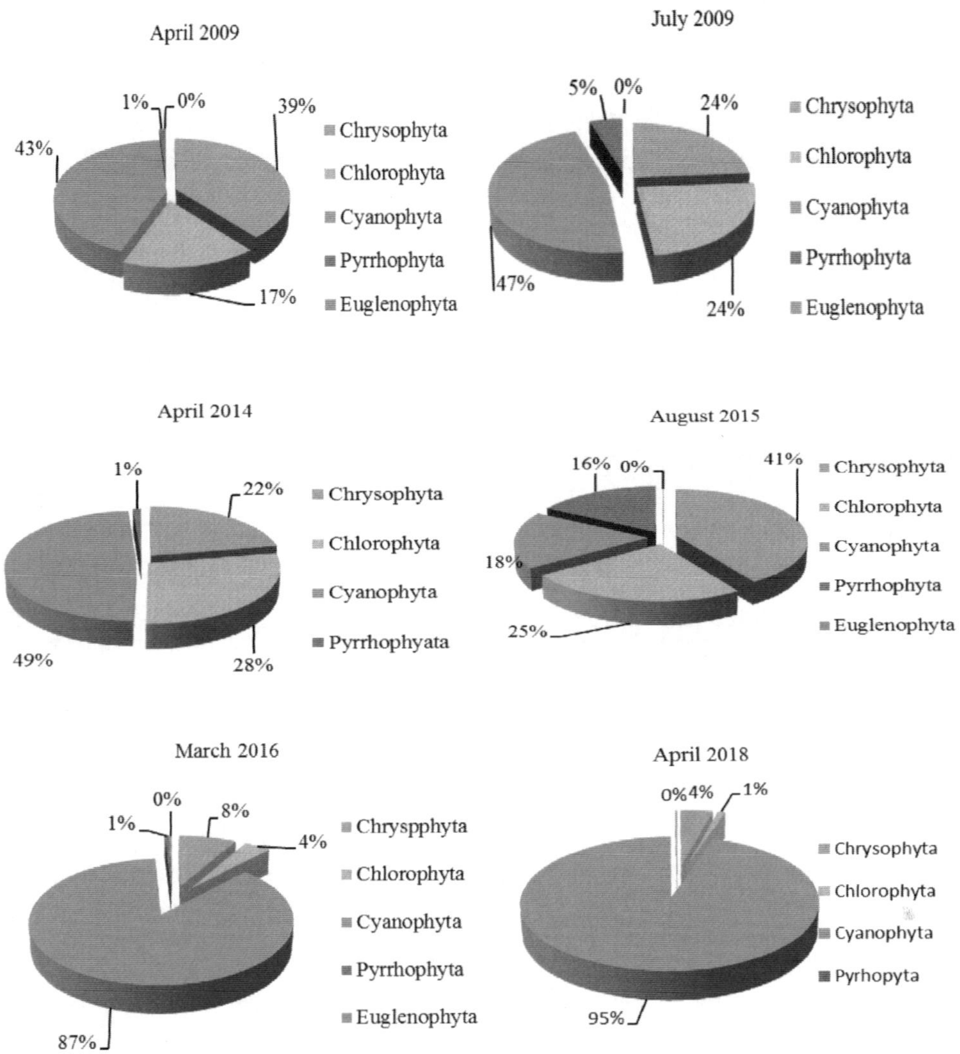

Figure 7. Phytoplankton composition in Lake Maninjau.

Table 3. Phytoplankton abundance of dominant species.

Dominant species	Phytoplankton abundance (x 10^3 individual. L^{-1})*					
	Non *Microcystis* bloom period			*Microcystis* bloom period		
	Apr 09	Jul 09	Apr14	Aug 15	Mar16	Apr18
Synedra ulna	4.95 – 36	65 – 88	14.2 – 103	12.3 – 13.18	0.088 -0.25	1.56 -1040
Monallantus sp	0	0	14 – 27	10.8 -23.25	0.05-11.09	0
St.playfairi	4.5-39.6	10.8-65.7	0.05 – 0.06	1.15 – 0.65	0 – 0.025	0
Aphanizomenon sp	17.1- 40.5	11.3- 54.5	0	0	0	0
A. affinis	4.5-21.6	7.2-57.6	0 – 0.015	0	0	0.5 -120
Chroococcus sp.	0	0	0 – 0.02	0 – 1.45	20.6 -77.67	0
C. raciborskii	4.05-6.75	16.2 – 81	3.55-8.5	2.53 – 4.075	0	0.0050 – 15
M. aeruginosa	0-1.8	0 – 0.09	0.005-0.005	0	0.024-0.01	37.9 – 24320
Planktolyngbya sp.	0	0	15.54 – 144	0 – 0.1	0	0
C. hirudinella	0	0	0	0.045-9.987	0	0

Remark * = Phytoplankton calculation in colony. Ap-09=April 2009; Jul-09= July 2009; Apr 14=April 2014; Aug15= August 2015; Mar16; March 2016; Apr18= April 2018.

3.4. Link between environmental factors and dominant phytoplankton.

Principal Component Analysis between phytoplankton and environmental factors showed that there were three components (figure 10). The first component was *Microcystis aeruginosa* and *Monnalantus* sp which was related to the environmental factor of temperature, pH, TN and TP for their dominance while the second component, was *Aphanizomenon* sp, *Anabaena affinis*, *Cylindrospermopsis raciborskii*, *Synedra ulna* which was related to conductivity and DO parameter for their dominance, the third group was *Ceratium hirudinella* which was related to TN:TP ratio for their dominance (figure 10, table 5). Pearson's correlation analysis showed that *Microcystis* had a positive correlation with TP, negative correlation with TN:TP ratios and Secchi depth transparency (table 5). It is indicated that increasing total phosphorus and lower TN:TP ratios create favourable condition for *Microcystis aeruginosa* bloom in Lake Maninjau. Among the physical and chemical factors, high TP concentration and low TN:TP ratio are most often used as indicators of cyanobacteria bloom. For instance, in Steilacoom Lake in Washington, blooms of *Microcystis aeruginosa* occurred when there was higher total phosphorus, decreased water transparency, high water column stability, high surface water temperature and pH [33]. And the success of *Microcystis* over other cyanobacteria was also associated with low nitrogen to phosphorus ratios [34]. In Lake Maninjau, water temperature and pH did not show a positive correlation with *Microcystis* bloom, however the average temperature (28.58 °C) and pH (7.75) are still in the range for *Microcystis* growth. In Bangladesh Lake, the initiation and persistence of *Microcystis* natural blooms was determined by relatively high water temperatures with the range of 28 to 30 °C [12]. Other study also reported that neutral and weakly alkaline conditions support the rapid growth of *M. aeruginosa* [35]. Based on this phenomenon, phosphorus was apparently the primary factors control *Microcystis* bloom in Lake Maninjau.

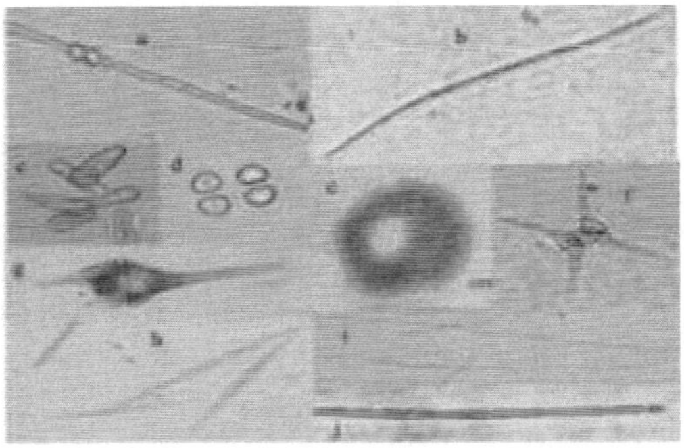

Figure 8. Photomicrographs of dominant species of phytoplankton in Lake Maninjau (a) *Anabaena affinis*, (b) *Aphanizomenon* sp., (c) *Monallantus* sp., (d) *Chroococcus* sp., (e) *Microcystis aeruginosa*, (f) *Staurasrum playfairi*, (g) *Ceratium hirudinella*, (h) *Synedra ulna*, (i) *Planktolyngbya* sp., (j). *Cylindrospermopsis raciborskii.*

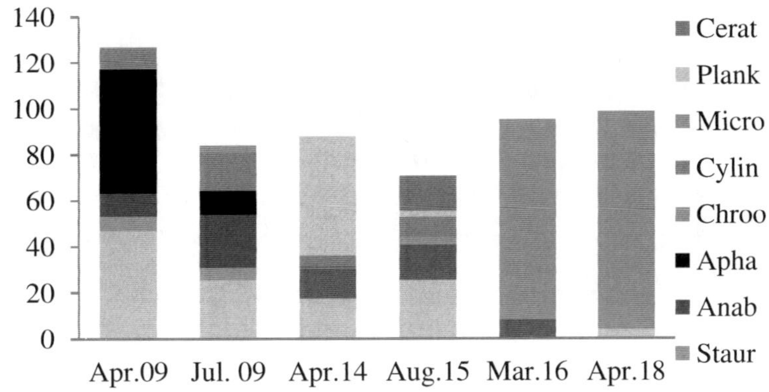

Figure 9. The percentage of phytoplankton composition of dominant species.
Remarks: Cerat=*Ceratium hiruninella*, Plank=*Planktolyngbya* sp.; Micro = *Microcystis aeruginosa*; Cylin=*Cylindrospermopsis raciborskii* Chroo=*Chroococcus* sp.; Apha=*Aphanizomenon* sp.; *Anab Anabaena affinis;* Staur =*Starastrum playfairi.;* Mol= *Monallantus* sp.; *Syn=Synedra ulna* sp.

Table 4. Pearson's correlation between phytoplankton and environmental factors.

	Temp	SD	pH	Con	DO	TN	TP	TN :TP
Synedra ulna	-0.12	-0.149	0.106	-0.388	-0.321	0.118	-0.263	0.183
Monallantus	**0.603**	**0.513**	0.246	0.342	0.35	**-0.606**	-0.322	0.253
Starurastrum	0.021	0.028	0.14	-0.413	**-0.583**	-0.048	-0.317	-0.018
Anabaena	-0.123	-0.203	0.017	**-0.665**	-0.264	0.082	-0.046	-0.28
Apanizomenon	-0.341	-0.199	0.133	-0.457	**-0.713**	**0.597**	-0.287	0.428
Chroooccus	**0.721**	**0.73**	**0.516**	-0.369	0.342	-0.185	-0.33	0.244
C. raciborskii	-0.082	-0.157	0.098	**-0.638**	-0.172	-0.042	-0.128	-0.236
Microcystis	-0.492	**-0.572**	**-0.725**	0.381	0.157	0.475	**0.87**	**-0.53**
Planktolyngbya	-0.222	-0.07	0.131	0.421	0.495	**-0.523**	0.009	-0.192
C. hirudinella	**0.291**	**0.166**	**-0.074**	**0.345**	**-0.311**	**-0.049**	**-0.248**	**0.461**

IOP Conf. Series: Earth and Environmental Science **380** (2019) 012020 doi:10.1088/1755-1315/380/1/012020

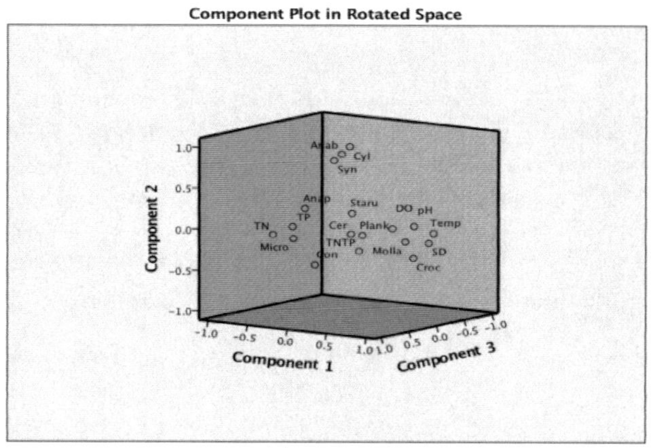

Figure 10. Component plot in rotated space from Principal Component Analysis between dominant species (*Anabaena*, *Aphanizomenon, Ceratium Chroococcus Cylindrospermopsis, Microcystis, Monallantu Planktolyngbia*, *Staurastrum*, *Synedra* and environmental factors) and environmental factors (Temperature, Secchi depth, Conductivity, DO, pH, TN, TP and TN:TP).

Table 5. Component matrix of dominant species and environmental factors.

Component Matrix [a]						
	Component					
	1	2	3	4	5	6
SD	0.899			-0.312		
Temp	0.895			-0.216	0.334	
TP	-0.789	-0.475	-0.135	-0.158	0.16	
pH	0.753	0.295	-0.164		-0.269	
Micro	-0.728	-0.412		-0.392	0.234	
Molla	0.659	-0.429	0.109	0.359	0.294	
TN	-0.587	0.288	0.551	-0.375		0.281
Anap	-0.213	0.786	0.317		-0.353	0.193
Con	-0.279	-0.782	0.338	0.37		-0.124
Anab	-0.267	0.752	-0.465		0.279	
Cyl	-0.172	0.731	-0.515	0.244	0.277	0.101
DO	0.155	-0.679	-0.633			0.302
Syn		0.658		0.503	0.228	0.391
TN:TP	0.529	0.298	0.681		-0.155	-0.195
Cer	0.284	-0.111	0.617	0.392	0.481	-0.103
Croc	0.657	-0.144	-0.101	-0.66		0.176
Plank		-0.42	-0.338	0.534	-0.589	
Staur		-0.338			-0.124	-0.774

Extraction Methods Principal Component Analysis
a: 6 component extract

Remarks:Anap=*Aphanizomenon* sp.; *Anab=Anabaena affinis;* Cer=*Ceratium hiruninella*.; Croc=*Chroococcus* sp.; Cyl=*Cylindrospermopsis raciborskii* Micro = *Microcystis aeruginosa;* Mol=*Monallantus* sp.; Plank=*Planktolyngbya* sp.; Staur=*Starastrum playfairi.; Syn=Synedra ulna* sp.

4. Conclusion

The most dominant assemblage in phytoplankton group in Lake Maninjau during cyanobacterial blooms was blue green algae. Phytoplankton species was dominated by *Microcystis aeruginosa* with a maximum abundance of 24,320 x 10^3 individual L^{-1} (94.4 % of the total assemblage). During the non-blooming period, cyanobacteria species were more diverse, represented by *Cylindrospermopsis raciborskii, Anabaea affinis*, *Aphanizomeon* sp., *Planktolyngbya* sp. and *Chroococcus* sp. Diatom (*Synedra ulna*) regularly occurred in all conditions, however, desmids (green algae) were disappeared during cyanobacterial blooms. It is highlighted that the occurrence of *Microcystis* blooms could be related to the total phosphorous dynamics in the lake.

5. References

[1] Paerl HW and Paul V 2012 *Wat. Res.* **46** 1349–63
[2] Brahmana and Bahri S 2002 *Buletin Pusair.Media informasi Kegiatan penelitian Keairan* **XI** (38) 23-28.
[3] Sulastri 2002 *Proc. Int. Symp. on Land Management and Biodiversity in South East Asia September 17 – 20 (Bali: Indonesia)* 403-8.
[4] Tanjung L R 2013 *OLDI* **39** (1) 30-8
[5] Syandri H, Junaidi, Azrita and Yunus T 2014 *J. Ecology and Environ Sci* **1** 109-13
[6] Junaidi, Syandri H and Azrita 2014 *J. Aquac. Res. Development* **5** 7
[7] Syandri H 2004 *J. Nature Indonesia* **6** 87-91
[8] Sulastri, Hartoto D and Yuniarti I 2012 *Ind. Fish. Res. J.* **18** (1) 1-12
[9] Nomosatryo S and Lukman 2011 *Limnotek*, **18**(2) 127-37
[10] Affan M A, El-Sayed Touliabah H, Al-Harbi SM, Abdulwassi N I, Turki AJ, Haque Md M, Khan S and Elbassat R A 2016 *Rend. Fis. Acc. Lincei.* **27** 473–81
[11] Duong T T, Jähnichen S, Quynh Le T P, Tu Ho C, Hoang T K, Nguyen T K, Vu T N, Kim Dang D K 2014 *Environ. Earth. Sci.* **71** 2419–27
[12] Marshall G H and Egerton TA 2009 *Va. J. Sci.* **60** (3) 149 –64
[13] Reynolds C S 1987 *Adv. Bot. Res.* **13** 67–143
[14] Harris G H 1986 *The phytoplankton Ecology. Sructure, Function and Fluctuation* (Chapman and Hall, New York) 384 pp
[15] An K G and Jones J R 2000 *Hydrobiologia* **432** 37–48
[16] Almanza V, Parra O, De M Bicudo C E, Baeza C, Beltran J, Figueroa R and Urrutia R 2016 *Revista Chilena de Historia Natural* **89** 8
[17] Fachrudin M, Wibowo H, Subehi L and Ridwansyah I 2002 *Proc. The 1st Seminar National Limnology 22 April 2002 (Bogor-Indonesia)* 65-75
[18] Apip 2004 *Warta Limnologi* **XVIII** (37) 7–10
[29] APHA 1999 *Standard methods for the examination of water and wastewater* 20th edition (American Public Health Association Washington)
[20] APHA 1975 *Standard methods for the examination of water and wastewater* 14th edition (American Public Health Association Washington) 1193 pp
[21] Prescott G W 1951 *Algae of the Western Great Lakes Area* (Crandbrook Institute of Science Bulletin No 31) 949 pp
[22] Scott A M and Prescott G W 1961 Indonesian Desmid. *Hydrobiologia* **XVII**
[23] Baker P D and Fabro L D 1999 *A guide to the identification of common blue green algae (Cyanoprokaryotes) in Australian Freshwater. Identification Guide* No 25 (Cooperative

Research Center for Ecology, CRC for Water Quality Treatment, Central Queensland University) 42 pp

[24] Gell P A, Soneman J A, Reid M A, Illman M A, and Sincock A J 1999 *An Illustrated Key to Common Diatom Genera from Southern Australia* (Cooperative Research Centre for Freshwater Ecology Australia) **36** 64 pp

[25] APHA 1992 *Standard Methods for the Examination of the Water and Waste Water* 17th Edition (APA-AWWA-WPCF) 1100 pp

[26] Santoso A B, Triwisesa E, Fakhrudin M, Harsono E and Rustini H A 2018 *IOP Conf. Series: Earth and Environmental Science* **118.** 012024

[27] Wetzel R G 1975 *Limnology* (W B Sounder Company, Toronto) p 60-65

[28] Hachitnson G E 1957 *Treatise on Limnology* **1** (Wiley and Sons Inc. New York) 1115 pp

[29] van Ginkel C E, Hohls B C and Vermaak E 2001 *Water S.A* **27**(2) 269-76

[30] Wiltsie D, Schnetzer A, Green J, Borgh M V and Fensin F 2018 *Toxins* **10** (92) 1-23

[31] Sulastri, Sulawesty F and Nomosatriyo S 2015 *OLDI* **41** (3) 339-53

[32] Wetzel R G 2001 *Limnology. Lake and River Ecosystem* (Academic Press New York, London) 1006 pp

[33] Jacoby JM, Collier DC, Welch EB, Hardy FJ and Crayton M 2000 *Can. J. Fish. Aquat. Sci.* **57**(1) 231-40

[34] Liu X, Lu X and Chen Y 2011 *Harmful Algae* **13** 37–43

[35] Yang J, Tang H, Zhang X, Zhu X, Huang Y and Yang Z 2018 *Environ. Sci. Pollut. Res.* **25** 4794–802

Acknowledgements

This research is funded by State Budget of Indonesian Government (DIPA-APBN and Priority National Research) through Research Centre for Limnology, Indonesian Institute of Sciences. We acknowledge RC for Limnology in providing the long-term monitoring limnological data of Lake Maninjau. The authors thank to Mrs. Rosidah for her assistance in field survey.

Bioaccumulation of Lead (Pb) in the macroalgae *Padina australis* Hauck in Makassar Marine Waters, South Sulawesi, Indonesia

Wahyu Supardi[1*] and Andhika Puspito Nugroho[1]

[1]Faculty of Biology, Universitas Gadjah Mada, Bulaksumur,Yogyakarta, Indonesia

**Corresponding author : wahyusupardi@outlook.com*

Abstract. Marine pollution is a global problem that occurs in coastal areas. Pb lead pollutants derived from waste discharges and Makassar community activities can accumulate on the macroalgae *Padina australis* Hauck. This study determined the accumulation of Pb on *P. australis* and assessed the abundance of *P. australis* in three stations in Makassar marine waters of Lae-lae, Samalona and Barranglompo from November-April 2018. Sampling *P. australis* was done by purposive sampling method. Data of *P. australis* abundance was taken using the line transect method. Physico-chemical parameters measured were temperature, salinity, pH, light penetration, current, DO, and BOD. Pb analysis was done using the Atomic Absorption Spectrophotometer. The results showed difference of Pb content in *P. australis* was found from each research station. The content of Pb at Station I is in the range of 0.0964 - 0.1388 ppm, Station II 0.0496 - 0.1050 ppm, and station III 0.0597 - 0.1035 ppm which has exceeded the water limit specified (> 0.008) ppm. The average concentration of Pb at station I was compared to other stations because Station I was closer to the mainland, with potentially greater entry of heavy metal Pb. Abundance of *P. australis* at station I were 92 ind/m^2, station II 152 ind/m^2 and station III were 319 ind/m^2. Correlation analysis data showed no significant value (ρ = 0.384, $p > 0.05$) which meant that the Pb concentration did not affect *P. australis* abundance.

1. Introduction

Marine pollution is a global problem faced by countries in coastal areas. The pollution is caused by the increasing construction of settlements, agriculture and industries that produced waste or pollutant containing waste. The entry of waste into the marine ecosystem can lead to decreased water quality. Decreasing the water quality can change the community structure. One of the major pollutants contained in waste disposal is heavy metals such as lead (Pb) and copper (Cu) [11].

One of the pollutant materials that could potentially endanger the coastal and marine areas of Makassar City is heavy metal lead (Pb). These metals are generally produced from transportation and industrial activities. Heavy metal Pb can accumulate in the macroalgae of *Padina australis* Hauck (Bioaccumulation). Bioaccumulation of heavy metals Pb by *P. australis* is strongly influenced by the high concentration of Pb metal in marine waters.

The coastal area of Makassar City is a densely populated urban area. Pb pollutants contaminating the waters of Makassar City comes from industrial activities such as paint, plastics, and other industries, ship building activities, ports, fisheries, hotels, marine tourism, and households producing marine discharges [22]. Furthermore, Hamzah (2007), stated that pollution in coastal waters of Makassar City is very high because there are two big rivers namely, Jeneberang and Tallo River, and canal or drainage from the city which flow to coastal waters of Makassar City, so that waters in Makassar potentially contain heavy metal such as lead (Pb).

The purpose of this study is to identify the level of accumulation of Pb on *P. australis* and to assess the abundance of macroalga *P. australis* in Makassar marine waters. Naturally, heavy metal Pb can enter the water through Pb crystallization in the air with the help of rain, in addition it also comes from the erosion of many mineral rocks around the water [16]. Organisms exposed to heavy metals Pb with low concentrations usually do not experience death, but will experience sublethal influence, that is the effect that occurs on the organism without causing death in the organism. This sublethal influence can be distinguished into three types, namely *inhibiting* (example growth and reproduction), causing morphological changes, and altering the behaviour of organisms.

Macroalgae has been shown to be efficient in metal monitoring and absorption [15]. *Phaeophyceae* has been shown to be effective as a heavy metal bioindicator because it has a high carboxylic density present in alginate [18]. The results show the analysis of lead concentration (Pb) in thallus of *P. australis* Hauck in Totok Bay waters by 3.8 ppm and in Blongko waters at 1.1 ppm. The concentration of lead (Pb) in sediment in Teluk Totok Bay is 22.6 ppm and in Blongko waters is 4.4 ppm. The concentration of lead (Pb) on thallus *P. australis* in Totok Bay waters and Blongko waters is above the standard quality standard of SNI No. 7387 Year 2009 of 0.5 ppm. The detection of lead concentration (Pb) in the thallus *P. australis* indicates a lead absorption process (Pb) is accumulated in the thallus [20].

Macroalgae *P. australis* can be found on all stations to be observed, therefore in this study the concentration of heavy metal lead in *P. australis* on water conditions in the three islands was examined.

2. Materials and methods

The materials used in this research are the samples of macroalgae *P. australis*, sediment samples in marine waters, seawater samples, formalin, manganese sulphate ($MnSO_4$), NaOH, sulfuric acid (H_2SO_4), Na-thiosulfate, nitric acid (HNO_3) , and aquadest. This research was conducted in November to April 2018. Observation and sampling were conducted in Makassar marine waters at three stations: station I in Lae-lae Island, II station in Samalona Island, and station III on Barrang Lompo Island. Sample preparation and analysis was conducted at Chemical Oceanography Laboratory, Faculty of Marine Science and Fishery, Hasanuddin University and Health Centre Laboratory, South Sulawesi Province using Atomic Absorption Spectrophotometer (AAS).

The subjects in this study included measurements of heavy metal lead (Pb) on samples *P. australis*, sea water and sediments on three water stations and the abundance of *P. australis* at three station in Makassar marine waters. Physico chemical parameters measured were temperature, salinity, light penetration, current, acidity degree, Biological Oxygen Demand (BOD) and Dissolved Oxygen (DO). The observations will be conducted on three islands; Lae-Lae Island, Samalona Island and Barrang Lompo Island to know the ratio of heavy metals in these three stations.

Figure 1. Sampling locations map.

Sampling of *P. australis* was done using the Purposive Sampling method. Each station is determined by 3 substations with sampling for analysis of 3 samples for 3 weeks with sampling interval 1 times a week. Number of samples of *P. australis* taken for metal analysis is 27 samples. Seawater and sediment samples were taken for each of 9 samples. Analysis of heavy metal Pb was done using the Atomic Absorption Spectrophotometer (AAS). The data collection of *P. australis* abundance was collected by line transect method. Each station was determined with 3 substations at intervals of 10 meters from each substation. Sampling was conducted on each substratum with square plot measuring 1 x 1 m. The total plots sampled were 45 plots. Physico chemical parameters measured were temperature, salinity, light penetration, current, acidity degree, Biological Oxygen Demand (BOD) and Dissolved Oxygen (DO).

Preparation of metal samples in macroalgae samples of *P. australis* Hauck were washed using distilled water and then dried in an oven at 120 °C until they reached a constant weight. The dried sample is then crushed into powder and homogenized with mortar and then stored in aluminium foil until the analysis is carried out. Next, the sample was weighed to a total of 1 gram, then ignited with a Furnace at 400 ° C for 20 hours until it became ash. Then, the ash is added with 4 ml concentrated HNO_3 + 1 ml concentrated HCl in a glass cup and heated using a hot plate. The filtrate is cooled at room temperature and placed in a 50 ml volumetric flask and dissolved with aquadest until the boundary mark, then filtered with Whatman 42 paper to obtain a clear yellowish solution ready for analysis.

Preparation of metal samples with 10 ml of sea water sample was put into Erlenmeyer flask, added with 0.5 mL concentrated HNO_3. The mixture is stirred until homogeneous, then heated using an electric bath until the solution dissolves. The dissolved solution was then put into a 100 mL measuring flask and added with distilled water. Samples are stored in plastic bottles ready for analysis. Preparation of metal samples in sediments were taken only on the surface of the substrate, namely the oxic part at a depth of 1-3 meters. Furthermore, in this section a measurement of lead metal concentration (Pb) will be carried out in the Laboratory.

Samples were also analysed to determine the Bioconcentration Factor (BCF) using the formula. Differences of Pb content in *P. australis* at three different stations were analysed using Two Way ANOVA with $p < 0.05$ indicating significance. Counting the population abundance of *P. australis*, Pearson

IOP Conf. Series: Earth and Environmental Science **380** (2019) 012021 doi:10.1088/1755-1315/380/1/012021

correlation analysis was performed on SPSS version 25 to determine correlation between metal concentration with *P. australis* abundance.

3. Results and discussion

3.1. Pb metal concentration in Padina australis

The result of the research indicate that there are differences in metal content of Pb at *P. australis* in every research station (figure 2). The content of metal Pb at Station I was in the range of 0.0964 - 0.1388 ppm, Station II 0.0496 - 0.1050 ppm, and station III 0.0597 – 0.1035 ppm which has exceeded the water limit specified that is > 0.008 ppm.

The average content of Pb concentration at station I was compared to other stations because Station I was closer to the mainland with potentially greater entry of heavy metal Pb in the sample. The high content of heavy metals in *P. australis* at station I is also caused by the low acidity of marine water with an average value of 7.25. The pH value is important to consider as it may affect the process and speed of chemical reactions in water. According to Phillips et al. (2009), *Phaeophyceae* has been shown to be effective as a heavy metal bioindicator because it has a high *carboxylic* density present in alginates (the main component of its cell wall). This can result in high levels of metal accumulation.

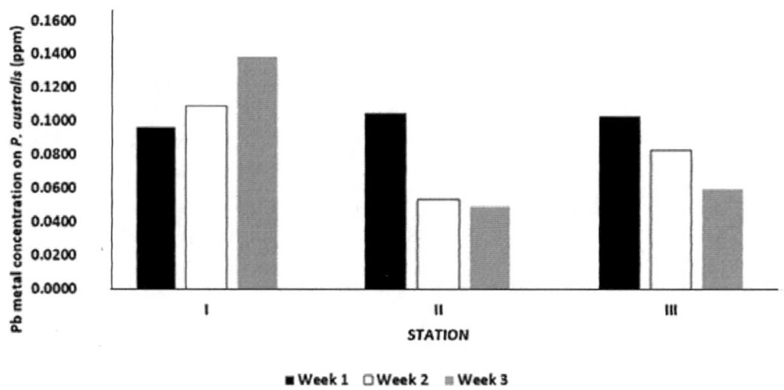

Figure 2. Pb metal concentration on *Padina australis* Hauck in each research station.

3.2. Pb metal concentration in sea water

The value of Pb metal concentration in sea water at station I is in the range of 0.0105 - 0.0172 ppm, station II in the range of 0.0059 - 0.0089 ppm and in station III in the range of 0.0062 - 0.0094 ppm (figure 3). The concentration of Pb metal between research stations is significantly different ($p < 0.05$), this is because the waters of Lae-lae Island (Station I) are closer to the mainland and the level of turbidity is very high. The heavy metal Pb that enters the aquatic environment will undergo deposition, dilution, and disperse, possibly absorbed by the organisms living in the waters [12].

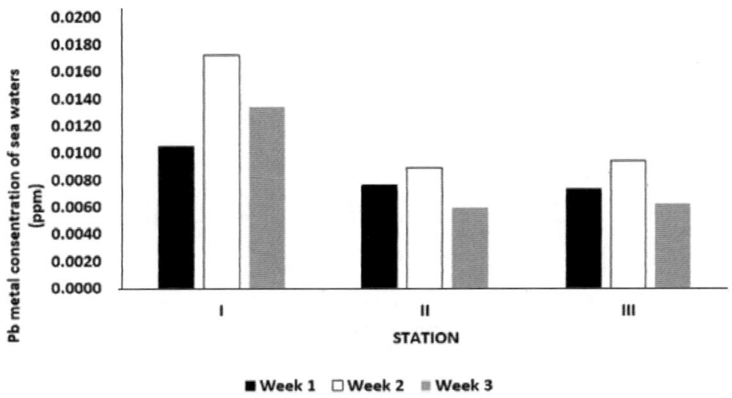

Figure 3. Pb metal concentration on sea water in each research station.

The average value of Pb metal concentration in the waters of Lae-lae Island (Station I), Samalona Island (Station II), and Barranglompo Island (Station III) has a value of Pb metal concentration that has exceeded the standard threshold of Pb metal content in marine waters i.e. > 0.008 ppm [5]. The highest concentration of Pb metal in the waters of Lae-lae Island (Station I), then Barranglompo Island (Station III) and next to Samalona Island (Station II). This is allegedly due to the close location of the waters of Pulau Lae-lae with the mainland which contributes a lot of waste both industrial waste, household and oil extraction from ships that anchored. The high content of heavy metal Pb in the waters of Barranglompo Island (Station III) compared to the waters of Samalona Island (Station II) was because Barranglompo Island is an inhabited island that has the potential to contribute waste to the water bodies.

3.3. Concentration of Pb Metal in sediment

The value of Pb metal concentration in Station I was in the range of 0.0293 - 0.0682 ppm, Station II in the range of 0.0072 - 0.0148 ppm, and Station III in the range of 0.0116-0.0278 ppm (figure 4). The metal concentration of Pb between research stations was significantly different ($P < 0.05$). At station I obtained the data of high metal concentration of Pb compared to station II and III. This is because the station I is closer to the mainland of Makassar city, so it is more likely to exposure to metal Pb in the sediment. Disposal of industrial wastes, household waste, and others, shipbuilding / docking, as well as oil drainage activities of vessels into the water bodies will occur in the sediment deposition process.

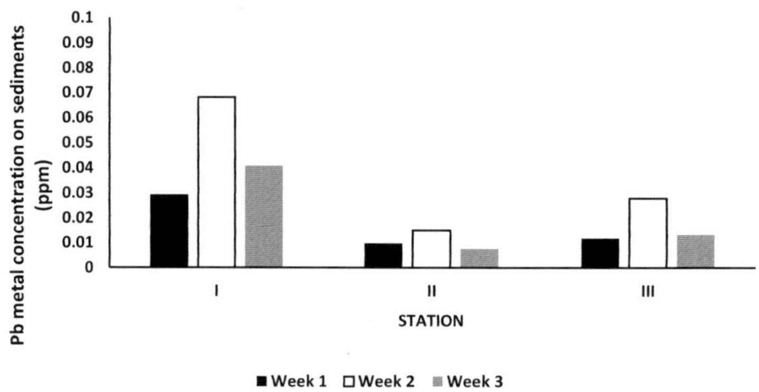

Figure 4. Pb metal concentration on sediments in each research station.

Pb metal can naturally enter the aquatic bodies through the complexation of metal particles in the air due to rain around the waters and human activities with various forms in which the contaminants will enter into the body of water and settle in the sediment [16]. Pb metal entering the aquatic environment will undergo deposition on the bottom of the water (sediments) and then it will be absorbed by the living organisms in the marine water [12].

3.4. Bioconcentration Factor

Bioconcentration factor is an ability of organisms to accumulate heavy metals from the environment. Lead accumulated in *Padina australis* probably originated from the water column. This happens because the entire thallus *Padina australis* was in the water column and all parts of the thallus are able to absorb heavy metals. The data obtained show that the Pb metal was absorbed by *Padina australis* very high that indicates *P. australis* has the ability to accumulate Pb metals. The Pb metal concentration found in *P. australis* affected the ability of Pb metal accumulation in *P. australis*. The concentration of Pb metal in *P. australis* at Station I was higher than the other stations (Station II and III) in the range of 0.0964 - 0.1388 ppm, so that the BCF value was lower than Station II and III because *P. australis* has low capacity in absorbing heavy metal Pb in station I (figure 5). The morphology of *P. australis* thallus found at Station I looks imperfect, with black coloured thallus and its intact shape which affects the ability of Pb metal accumulation in *P. australis*. Station III has a high BCF value than stations I and II because at station III the environment is good with suitable and less polluted environmental conditions, which supports *P. australis* so that *P. australis* has the ability to accumulate heavy metals well in the thallus.

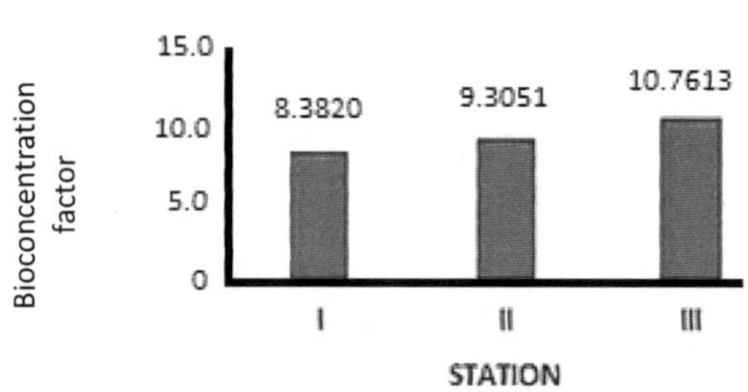

Figure 5. Bioconcentration factor value.

Janssen et al., (1997) stated that if the BCFs value > 1, it means that the organism has the ability to accumulate Pb metal in its body, whereas BCFs ≤ 1 means that the organism lacks or does not have the ability to accumulate metals in its body.

3.5. Abundance of macroalgae P.australis Hauck

Abundance of *P. australis* at Station I was 92 Ind / m^2, Station II 152 Ind / m^2, and Station III 319 Ind / m^2 (figure 6). The result of Pearson correlation analysis data showed value of 0.384, $p > 0.05$, that is there is no correlation between Pb concentration with abundance, but the direction of relation shows negative number (-0.331) which could indicate that if Pb concentration rises then *P. australis* abundance will decrease.

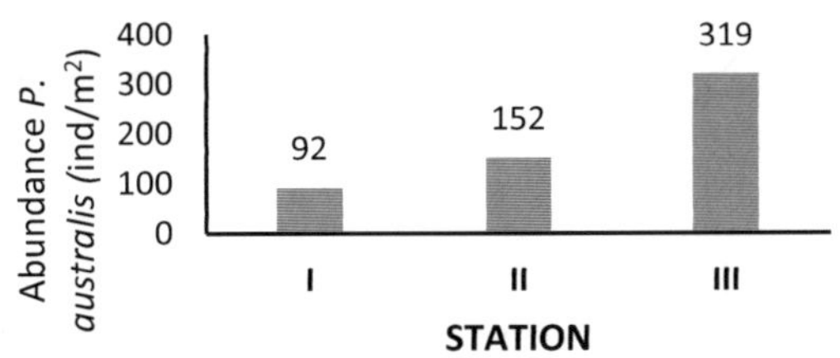

Figure 6. Abundance of macroalgae *P. australis* at the research station.

3.6. Physico-Chemical Parameters

From observation, mean value water temperature at the research station is 28.67 - 29.33 ºC (table 1). The value still support algae growth that is 15 - 30º C for the tropics. The average salinity value is 29 - 34 ‰ (table 1). The optimum salinity conditions for macroalgae plants range from 25 - 35 ‰. The pH measurement results in the three stations ranged from 7.25 - 7.36 (table 1). The pH value in the range of values is still feasible for algal growth. According to Odum (1971) a good pH for algal growth is 5 - 8. Average current velocity values range from 0.089 - 0.220 m/s (table 1). The average value of sea water light penetration ranges from 1.62 to 4.86 m. The standard of sea water quality according to the Minister of Environment No. 51 of 2004 is > 5 (table 1).

Table 1. Measurement of chemical physiological parameters based on seawater quality standards for marine biota (mean ± STDEV).

No.	Physico Chemical Parameters	UUnit	Station I (Lae-lae Island)	Station II (Samalona Island)	Station III (Barranglompo Island)	Quality standards*
1	Temperature	ºC	28.67 ± 0.01	29.33±0.01	29±0.00	28-30
2	Salinity	‰	29.00 ± 2.65	33.67±0.58	34.00±1,00	33-34
3	pH	-	7.25 ± 0.03	7.33±0.03	7.36±0,08	7-8.5
4	Current velocity	m/s	0.0089 ± 0.012	0.160±0.016	0.220±0.029	-
5	Light penetration	m	1.62 ± 0.38	4.17±1.16	4.86±0.23	>3
6	DO	ppm	3.51 ± 0.53	6.42±0.25	6.64±0.16	>5
7	BOD	ppm	2.74 ± 0.42	1.38±0.42	1.32±0. 41	<10

The condition of the waters of Station I is classified as moderately contaminated with the value of 3.51 ppm, station II 6.42 ppm and station III is classified as light contaminated category, 6.64 ppm. The threshold value or seawater quality standard according to the Minister of Environment No. 51 of 2004 is DO > 5. The average value of BOD obtained was in the range of 1.32 to 2.74 ppm. The result of BOD5 obtained is <10 ppm, as according to Effendi (2003) if the waters are characterized by BOD5 of more than 10 ppm then this indicates the area is contaminated.

4. Conclusion

The content of metal Pb on macroalgae of *P. australis* Hauck at Station I is in the range 0.0964 - 0.1388 ppm, Station II 0.0496 - 0.1050 ppm, and station III 0.0597-0.1035 ppm which has exceeded standard defined water limit is > 0.008 ppm. The average height of Pb concentration at station I was compared to other stations because Station I was closer to the mainland with potentially greater entry of heavy metal Pb in the samples. Abundance of *P. australis* at station I were 92 individuals, station II 152 individuals and station III 319 individuals. The result of correlation analysis data showed no significant value ($\rho = 0.384$, $p > 0.05$) which mean that the Pb concentration did not affected *P. australis* abundance.

5. References

[1] Akcali I, and F Kucuksezgin 2011 A biomonitoring study: Heavy metals in macroalgae from Eastern Aegean coastal areas *Marine Pollution Bulletin.* **62** 637 - 645

[2] Atmadja W S 1996 Introduction of Red Algae Type. In: Introduction of Indonesian Seaweed Types *Center for Oceanographic Research and Development* Indonesian Institute of Sciences Jakarta

[3] Bryan G W 1976 Some aspects of heavy metal tolerance in aquatic organism. In: A. P. M. Lockwood (ed.*) Effects of polltitants on aquatic organisms* Cambridge England

[4] Connel D W, and J G Miller 1995 *Chemical and Ecotoxicology Pollution* University of Indonesia Press

[5] Decree of the Minister of Environment Number 51 of 2004 on the Quality Standard of Sea Water

[6] Doty M S 1985 *Eucheuma alvarezii sp.nov (Gigartinales, Rhodophyta) from Malaysia. In: Abbot I.A. and J.N. Norris (editors) Taxonomy of Economic Seaweeds California Sea Grant College* Program pp 37 - 45

[7] Doty M S 1986 *Biotechnological and Economic Approaches to Industrial Development Based on Marine Algae in Indonesia Workshop on Marine Algae Biotechnology. Summary Report* (England: National Academic Press Washington DC) pp 31-34.

[8] Filote C, G Ungureanu, R Boaventura, S Santos, I Volf and C Botelho 2016 Green macroalgae from the Romanian coast of Black Sea: Physico-chemical characterization and future perspectives on their use as metal anions biosorbents. *Process Safety and Environmental Protection* http://dx.doi.org/10.1016/j.psep.2016.06.002

[9] Febries and G Warner 1994 *Characterization of toxicants is sediments from Part Philip Bay: Metals.* (Australia: Departement of Conservation and Metals Resources Melbourne)

[10] National Standardization Agency 2009 *The maximum limit of heavy metal contamination in food.*

[11] Hastuty 2008 *The content of heavy metal lead (Pb) in Sargassum sp. as a pollutant bioindicator in Supermonde Islands* [thesis]

[12] Hutagalung H P, and A Rozak 1997 *Determination of Nitrate Level. Methods Analysis of Sea Water, Sediment, and Biota.* H. P Hutagalung, D. Everyermana and S. H. Riyono (Editor) (Jakarta: Center for Oceanographic Research and Development LIPI)

[13] Kramer K J M, U H Brockmann and R M Warwick 1994 *Tidal Estuaries: Manual of sampling and Analytical Procedures* (Rotterdam: A A Balkema)

[14] Lee S H and Park CH 2012 Biosorption of heavy metal ions by brown seaweeds from southern coast of Korea *Biotechnol Bioprocess Eng.* **17** 853-861

[15] Montazer-Rahmati MM, Rabbani P, Abdolali A, Keshtkar A R 2011 Kinetics and equilibrium studies on biosorption of cadmium, lead, and nickel ions fr aqueous solutions by intact and chemically modified brown algae *J Hazard Mater.* **185** 401–407

[16] Palar H 1994 *Heavy Metal Pollution and Toxicology* (Jakarta: Rineka Cipta)

[17] Palallo A 2013 *The distribution of macroalgae in the ecosystem of seagrass and coral reefs in Bonebatang Island, Ujung subdistrict, Barrang lompo village, Makassar* Department of Marine Sciences, Faculty of Marine Science and Fisheries, Hasanuddin University

[18] Phillips D J H 1990 *Use of macroalgae and invertebrates as monitors of metal levels in estuaries and coastal waters. In: Furness RW, Rainbow PS (eds) Heavy metals in the marine environment.* CRC Press, Boca Raton pp 81-99

[19] Praveena, Jasmine R and Estherlydia D 2014 Comparative Study of Phytochemical Screening and Antioxidant Capacities of Vinegar Made From Peel and Fruit Of Pineapple (*Ananas comosus* L.) Food Chemistry and Food Processing, Loyola College, Chennai. *International Journal of Pharma and Bio Sciences.* **5(4)** 394-403

[20] Siahaan, B, Desy M H M and Joice R 2017 *Lead Metal Analysis (Pb) and Chlorophyll concentration on Algae Padina australis Hauck From Totok Bay Waters.*

[21] Sudding S and Dewi A 2012 Analysis of Lead (Pb) levels at the root of white fire (*A. alba*) at the Jongaya sewerage on Jalan Metro Tanjung Bunga Kota Makassar *Journal of Chemica* **13** State University Makassar

[22] Sheehan, J, T Dunahay, J Benemann, and P Roessler 1998 *A look Back at The U.S. Departmentof Energy's Aquatic Species Program : Biodiesel from Algae* (Colorado)

[23] SNI Field of Public Works Regarding WATER QUALITY 1990 Edition SK SNI M - 49-1990 03 *Ministry of Public Works*

[24] Wood D, E Capuzzo, D Kirby, K M McAuley and P Kerrison 2017 UK Macroalgae aquaculture: Whay are the key environmental and licensing considerations ? *Marine policy.* **83** 29-3

Acknowledgments
The author would like to thank the Bioskeleton UNHAS 2011, Biological Celebes Diving Club and South Sulawesi Divers for provided considerable support in this research process, both at the time of survey and sampling in the Field. Makassar Centre for Health Laboratory, and Chemical Oceanography Laboratory, Hasanuddin University, South Sulawesi which has helped provide and provide laboratory facilities.

AUTHOR INDEX

Institute of Physics
Dirac House, Temple Back
Bristol BS1 6BE UK

ISSN: 1755-1307
ISBN 978-1-7138-1300-2